FALLEN

A TRAUMA,
A MARRIAGE, AND THE
TRANSFORMATIVE
POWER OF MUSIC

FALLEN

KARA STANLEY

GREYSTONE BOOKS
Vancouver/Berkeley

Greystone Books Ltd.
www.greystonebooks.com

Cataloging data available from Library and Archives Canada
ISBN 978-1-77164-102-9 (pbk.)
ISBN 978-1-77164-103-6 (epub)

Editing by Nancy Flight
Copyediting by Stephanie Fysh
Cover design by Peter Cocking
Text design by Peter Cocking and Nayeli Jimenez
Printed and bound in Canada by Friesens
Distributed in the U.S. by Publishers Group West

Diagram on p. 71 by Nayeli Jimenez
An adaptation of a cortical localization diagram on p. 122 is printed
with the permission of the Centre for Neuroskills, neuroskills.com,
using art from shutterstock.com.

We gratefully acknowledge the financial support of the Canada
Council for the Arts, the British Columbia Arts Council, the Prov-
ince of British Columbia through the Book Publishing Tax Credit,
and the Government of Canada through the Canada Book Fund
for our publishing activities.

Greystone Books is committed to reducing the consumption of
old-growth forests in the books it publishes. This book is one step
toward that goal.

We need only view a Dissection of that large Mass, the Brain, to have ground to bewail our Ignorance. On the very Surface you see varieties which deserve your admiration: but when you look into its inner Substance you are utterly in the dark, being able to say nothing more than there are two Substances, one greyish and the other white, which last is continuous with the Nerves distributed all over the body . . . If this Substance is everywhere Fibrous, as it appears in many places to be, you must own that these Fibres are disposed in the most artful manner; since all the diversity of our Sensations and Motions depend upon them. We admire the contrivance of the Fibres of every Muscle, and ought still more to admire their disposition in the Brain, where an infinite number of them contained in a very small Space, do each execute their particular Offices without confusion or disorder.

FROM SEVENTEENTH-CENTURY DANISH PHYSICIAN, GEOLOGIST, AND PRIEST NIELS STENSEN'S *DISCOURS SUR L'ANATOMIE DU CERVEAU*

CONTENTS

NOTE TO THE READER

DR. CHARLES SUNG Haw, Dr. Donald Griesdale, Dr. Marcel Dvorak, Dr. Jennifer Yao, Dr. Rhonda Willms, and the members of Simon's rehabilitation team—Sean, Melissa, Natalia, and Richard—have all kindly agreed to appear in this memoir as themselves. Due to the difficulty of contacting all the doctors, nurses, and therapists, many of whom we were only on a first-name basis with, I have, in all other instances, changed the names of the various medical professionals we encountered in order to protect their privacy.

RUGGED CHEDDAR AND THOUGHT-PRAYERS

11:20 A.M., JULY 22, 2008

RUGGED CHEDDAR. THIS is what I am thinking about when my husband, Simon, falls. I have been thinking about cheese, and about Simon, on and off all morning. The deli counter at the grocery store has set out samples of a particularly good aged cheddar, and as a small square dissolves on my tongue, the crunch of tiny lactic acid crystals sweetening the salty, nippy kick, I decide to blow our weekly budget and buy a block. Simon will love it. If a man could claim a cheese of his own, rugged cheddar would belong to him. Over the years we have had disagreements about cheese consumption. I have had occasion, during particularly tight financial times, to assert that cheese is a *meal* item. It is not a snack. I have wailed as I watched Simon slice into thick slabs and devour what I considered two nights' worth of dinner material in an attempt to restabilize a late-afternoon dip in blood sugar. But thankfully, those times have passed. Although we still live paycheck to paycheck, there is enough buffer room in our budget to accommodate a totally superfluous block of cheese.

I will cut it up and serve it with a plate of red peppers, and open a bottle of wine. Even though there is no special occasion to mark, I will surprise Simon with a date-night dinner. He was tired this morning when he woke, and achy and anxious about the coming weekend. Eli, our sixteen-year-old son, is away at soccer camp, and we have only just made it through a long and fractious winter. Damp and gray, the cold months lingered deep into spring, and the grind of extra-long work days, pressing deadlines, and a sense of stalled forward motion wore us both down. There is need, and now some time, for a few spontaneous date nights.

Earlier this morning I opted to ignore the stack of editing work piled beside my computer and instead took an early-morning yoga class, my first in over a year. After yoga, I moved through a list of chores at a meditative pace, as if this were my full-time job, work to take pride and pleasure in instead of something usually squeezed in at the end of a long day, tasks turned unpleasant by the gnawing sense that there is never enough time to get things fully done. Because I know how tired Simon is, how deserving of a break, the leisurely pace at which my day unfolds is a guilty pleasure.

Still, the sun's warmth creates space, opening up the hours, and I soften into a state of happy anticipation. It's too hot to eat a heavy meal, so the dinner I plan takes the shape of a table laden with appetizers: a dish each of Kalamata olives and marinated artichokes, sliced avocado, cantaloupe, and garden-fresh greens. A warm baguette. I'll make a pot of soup for a main course—a minestrone, perhaps. Or a smooth tomato and basil, Simon's favorite. It will be a meal of simple luxuries, one to savor long into the late sunset of a summer evening. This is what I am thinking: rugged cheddar, a bottle of red wine, and an evening alone with Simon. It is 11:20 a.m., and I am in my car driving home to Halfmoon Bay.

Back at home, as I unpack the groceries, the warm blue buzz of summer air is fractured by a blistering series of sirens: ambulance, police car. Another police car? It must be a serious accident. In a community as small as Halfmoon Bay, it is likely that even if you don't know the person in the ambulance, you know someone who knows them. As I always do, I send out a thought-prayer of strength and safety to whomever the ambulance is intended for.

Be safe.

Be strong.

There is a moment of silence in the wake of the sirens' keening trill, but as I return to my average, if exceptionally leisurely and pleasurable, Tuesday, everyday noises resume: the frantic chatter of insects, the lush and liquid warble of birdsong, kids shouting, a chainsaw's electric purr somewhere in the distance. I continue to catch up on household chores. I attend to the recycling, the pileup of dirty clothes, the grimy bathroom. I fill a vase with fresh lavender and a single blood-red rose and place it on the bedside table in our bedroom, then change into my dirt-streaked gardening shirt in preparation for starting some much-needed yard work. I am fitting clean sheets onto the bed before heading outside when there is a knock at my kitchen door.

The knock surprises me. The dog starts barking like a maniac, and I abandon the fourth fitted bedsheet corner to answer the door. Framed by the kitchen window is the distinctive profile of our friend Ryan: his long, dark hair pulled into a tight bun at the nape of his neck, his Daliesque mustache in a particularly stiff twist.

"Come on in," I shout as I reach for the kettle. Simon and Ryan are building a house down the road on a large waterfront property, working longer hours than usual to complete as much of the project as possible by the end of this week. I fill the kettle with tap water and switch it on. Maybe they're home for a late lunch. But why the knock? And where is Simon?

It is Lou, one of Simon's bosses, who steps through the door, and I know then that something is wrong. My body knows: an icy chill bristles the back of my neck while deep in my belly a seasick clutch of panic surges and swells. My heart, an unmoored boat in an unexpected storm, capsizes and beats painfully against my ribs. I can't breathe.

Lou speaks in short sentences. "Simon fell. He fell on his back." He takes a deep breath and then continues in the same calm, measured voice. "He couldn't feel his legs after his fall. Don't worry—that's not uncommon. It is shock. An Air Evac landed at Trout Lake, and they've taken Simon to Vancouver General Hospital. This is protocol. This is precautionary. We're just being safe."

For a moment all I can do is picture Simon, earlier this morning, as he stood at the foot of the bed, shirtless and scowling, rubbing his wrist.

It was sore.

It was sore enough that he considered staying home, a rare occurrence. He strained it, he figured, playing his slide arrangement of the Paul Simon song "Diamonds on the Soles of Her Shoes" Sunday night, and he was worried that it wouldn't be in top playing form for his weekend gig. But there was a deadline to meet, a house to shingle. This was his last week before an extended break to focus on music. Besides—a sore wrist? In the macho culture of construction workers, that's a pretty thin reason to take a day off. Weak.

"It hurts," he said. "But it's my last week. Shit." He wavered, not only irritable, but also unusually indecisive.

"He wasn't going to go to work today," I say now to Lou, as if this is a critical piece of information. I look to Ryan, sitting at the kitchen table, but he doesn't say anything. I repeat myself. "He wasn't going to go to work today."

"I wish he hadn't," Lou says. He outlines a plan: Ryan will drive me to the ferry to Vancouver and then to the hospital. Lou and Dave, the owner of the construction company, will follow us in a second vehicle. I should pack an overnight bag, clean clothes for Simon.

There are things to do: I call the Sunshine Coast Association for Community Living, where I work part-time, and ask them to cover my week's shifts. I call my mother in Powell River and briefly update her. I change back into a clean T-shirt and dig in the closet for some kind of overnight bag. When I don't find one there, I drift from room to room, lost. Lou continues to give me prompts. "Maybe pack a pair of sunglasses," he suggests. "Simon might have a concussion; his eyes might be sensitive to the sunlight."

Lou is a volunteer fireman and has some experience handling crises. It dawns on me that he's here partly because he has this experience. Right now, I am the situation he is handling. This thought—more than what he is saying—makes me feel, in quick succession, resentful, appreciative, and, finally, afraid. "The dog?" I say.

"Don't worry," Lou says. "We'll take care of her."

Duffel bag found, I haphazardly fill it with a change of clothes for both Simon and me, a bent and scratched pair of old sunglasses, and the Cormac McCarthy novel—*All the Pretty Horses*—that Simon is reading. If he has to spend a few days in hospital, he will be happy to have his book. As I continue to pack odds and ends, I calculate the time it will take to reach Vancouver General Hospital. The Sunshine Coast is connected to the Lower Mainland by a forty-five-minute ferry ride. The 1:30 ferry will just be pulling out now; the next sailing won't be until 3:30. Then it will take at least an hour, depending on traffic, to reach the hospital. Another seasick swell and surge, the waves growing bigger. I steady myself against the bathroom counter as I add up the hours: I won't be with Simon until almost 5:30.

"When?" I ask Lou, stuffing toothbrushes and contact lens solution into the duffel bag. "When did this happen?"

"Just after the morning break," he says from his post at the kitchen door. "About 11:20."

"I heard the sirens," I say as I pass through the kitchen, grabbing the phone and returning to our bedroom. This too feels like a critical piece of information. I heard the sirens. I sent out a thought-prayer. But ... that was an hour and a half ago. Why hadn't they called me then? I could have met the Air Evac with Simon, or at least been on the 1:30 ferry. I don't ask this out loud, recognizing it as the kind of maternal questioning that drives my son crazy. There's no use second-guessing actions already taken, Eli would say if he were here now, and he'd be right. I call our oldest friend, Guido, and leave a message: Could he meet Simon at the hospital? He calls back moments later. He will get to the hospital as fast as he can.

As I hang up, the phone rings. Vancouver General Hospital. On the other end of the line the nurse, like Lou, speaks in slow, measured sentences: Upon Simon's arrival at the hospital, his left pupil was fixed and dilated. They intubated him and the condition of his eye improved. They would have to operate. When would I be there?

Lou is visibly shaken when I relay this information. "Simon was totally lucid," he says. "He was talking normally. He vomited and there was some blood in his ear, but he was talking totally normally. I told him I'd get you and bring you to the hospital. He got a little angry, swearing, when they loaded into the helicopter but ..." Lou's voice trails off and he clasps his hands together. "I should go get Dave. We'll meet you at the ferry."

Unlike Lou, I am momentarily grounded by the news from the hospital. But what I choose to hear is selective. In my mind, Simon arrived at the hospital in serious condition and they had to intubate, but that is over and he is improving now.

"Improving"—the only word that mattered. I consider calling Eli and arranging to have him meet me in Vancouver, and also calling Simon's parents, Marc and Lorna, in Quebec, but I don't.

Instead, I throw my copy of Joan Didion's *The Year of Magical Thinking* into the duffel bag. Joan Didion is at the top of the list of authors I hold dear as any friend and consider a worthy traveling companion in a time of crisis. I have already revisited the book several times during the last year in the wake of my mother's diagnosis of breast and lymphatic cancer, when I was in need of Didion's insight, intelligence, and honesty. I stared at the author's photograph on the back flap as if I could locate the source of her power in her ungrayed hair, birdlike bones, and pale, almost translucent skin. I worried about her when I learned, postpublication, of her daughter's death. I wanted to bring her soup. To return, in some small way, the kind of comfort she had provided me over the years.

The Year of Magical Thinking is Didion's account of the year following her husband's sudden death and her daughter's near death from an inexplicable flu turned full-body sepsis and resulting bleed in the brain. In the book, Didion relates how she was told several times, as evidence of the severity of the situation, that en route to surgery one of her daughter's eyes was "blown," meaning it was fixed, dilated, and nonreactive to light. A fixed and dilated pupil—an FDP—signified compression of the brain stem, which in turn signified a shutting down of the most basic and critical bodily functions: heartbeat, blood pressure, breath. An FDP signified brain death.

Even though I can't consciously recall it, the facts Didion quotes from a study done by the Department of Neurosurgery at Bonn's University Clinic are embedded somewhere in my mind: "The study followed ninety-nine patients who... presented with... one or two FDP's. The overall mortality rate was

75 percent. Of the 25 percent who were still alive twenty-four months later, 15 percent had what the Glasgow Outcome Scale defined as an 'unfavorable outcome' and 10 percent a 'favorable outcome.' " Didion goes on to translate the percentages: "... Of the ninety-nine patients, seventy-four died. Of the surviving twenty-five, at the end of two years, five were vegetative, ten were severely disabled, eight were independent and two had made a full recovery."

So even though I tell myself that I don't want to worry Marc and Lorna and Eli needlessly, that I will call when I have more real information, I have a pretty clear understanding of the significance of a fixed and dilated pupil. This, then, is the beginning of my magical thinking. I can't call Simon's mother and father or our son because that would mean this is all really happening.

Bag packed, I say good-bye to the dog, Paloma. I tell her to be brave and good and not to bark too much at whoever comes to feed her. I tell her Simon and I will be home soon, hoping that with the simple act of saying it out loud I can make it true.

Ryan and I leave for Vancouver.

{ 2 }

A HUNGRY BEAST

T HE AVERAGE ADULT brain weighs approximately three pounds. Of all the mammals, humans have the highest brain-to-body-mass ratio, though when it comes to the human mind it is not clear how much size matters. Dmitri Mendeleev, the chemist who created the periodic table of elements, had a brain that was slightly larger than average, and Russian novelist Ivan Turgenev's brain was a whopping 50 percent larger than the norm; yet the sacred tissue of Einstein's brain was a little shy of the three-pound standard. It has been demonstrated that the more years a London taxi driver has spent driving the complex weave of London streets, the greater the size of the hippocampus, the region of the brain that stores spatial representations—strong evidence that our daily activities shape and build various structures of the brain.

Likened sometimes to a bowl of curds, the brain has the consistency of a thick gelatin or a block of tofu and is held together by three layers of membranes: the dura mater, or "tough mother," which is the thick, outermost layer; the arachnoid mater, or "spider mother," a delicate layer attached to the dura; and the pia mater, the "gentle mother," which is the thin, fibrous innermost layer. All float in a sea of cerebrospinal fluid, a clear, colorless liquid composed mostly of water, which circulates through the gaps and cavities of the brain, cushioning and protecting the delicate neural tissue from the internal bony ridges of the skull. At the floor of the skull is a small opening, the foramen magnum, where the brain stem exits the cranial cavity and carries messages back and forth between the brain and the spinal cord and from there to the whole of the nervous system and the rest of the body.

The soft substance of the brain is commonly referred to as gray matter and white matter, terms that describe the color of tissue when it has been removed from a body, fixed and hardened by preservatives. In a living body, the outer core, or gray matter, of the brain is actually a tawny pink, and the inner core is an off-white blush. This living tissue comprises roughly 100 billion neurons, or nerve cells, and about 10 to 50 trillion neuroglia, the "glue" cells that support, nourish, and protect the neurons. Each neuron can form roughly one thousand synapses, or connections, with other neurons, making the overall number of potential synaptic connections of the average human brain upward of a thousand trillion, a number greater than the combined population of people on our planet and stars in our galaxy. This vast neural circuitry, faster and far more immense than any machine ever assembled by human hands, is both a wonder and a mystery, constituting who and what and how we are in the world. Not only is our neural network responsible for how we calculate a complex math equation, execute the fine motor skills needed to perform brain surgery, or memorize a musical score note for note, it also determines such character traits as aggression, jealousy, and empathy; it influences the dynamics of our personality and the perimeters of our sense of self, and determines who and how we love. And it creates our consciousness, determining, in effect, how we think about how we think.

This immensely powerful neural network is also breathtakingly fragile, for the neurons of the central nervous system have little or no capacity to replicate or repair themselves. Both bleeding and swelling in the brain can cause neural tissue to become necrotic and die, as can lack of oxygen. Although the brain constitutes only 2 percent of the average body's weight, it is a hungry beast that consumes 20 percent of the oxygen circulated in the blood. If the supply of oxygenated blood is interrupted for even two or three minutes, brain cells begin to die.

{ 3 }

WALKING BLUES

I AM GUT-SICK WITH worry about how worried Simon must be, alone in an air ambulance possibly unable to feel his legs. *I'm coming, Beau, I'm coming.* I will my new thought-prayer to travel the distance between us. *I'm coming. Don't worry. I'm coming.* But the slow pace of ferry travel is agonizing. Ryan and I wait for what feels like forever to board the boat—which is forty minutes behind schedule—and finally, as the horn sounds and we slowly pull away from the dock, I curse my lack of a bridge or speedboat or wings—anything that would get me to Simon faster.

Up on board, Ryan and I find Lou and Dave, the owner of the construction company, and the four of us sit in a stricken silence amid the silvery spangle of sunlight meeting ocean and the hoopla of families embarking on summer vacations. Gradually, as each of Simon's co-workers offers a new bit of information about the morning's events, a moving picture, in all its horrific detail, starts to unreel in my mind.

Simon. At work, not far from our house, building a multi-million-dollar waterfront home. He swallows a last sip of coffee, finishing his mid-morning break, and returns to the second-story scaffolding, where he continues installing cedar siding on the outside of the house. It is 11:20 when he pauses the swing of his hammer and steps backward. The board tips under his feet and gives way. Simon reaches, grabs air, and falls, the loose section of scaffold twisting in the air until it slices into the ground beside him, narrowly missing severing his body in two. He lands on his back: skull and spine meeting hard stone tile.

"I can't feel my legs," he says and sits up. Ryan rushes over as Simon removes his work belt before falling back to the ground. Blood drips from his left ear, and he vomits onto Ryan's white T-shirt. He repeats himself: "I can't feel my legs."

When Lou, a volunteer fireman, is told of Simon's condition, he calls to notify the paramedics who are on their way that an Air Evac is required immediately.

I can picture it, but I can't really understand it. My mother calls Dave's cell to tell me she has talked to Vancouver General Hospital.

"Oh, honey," she says, "it's serious. There's bleeding in the brain. I'm going to get on a plane and meet you in Vancouver tonight."

"No, don't," I tell her. "Wait. I'll call you from the hospital." As if by saying no to my mother I can say no to a bleed in the brain. I caution myself not to panic. I will wait and speak to the doctors myself. My mother has always been panicky, a social worker prone to see the worst-case scenario. It was my mother who told Simon over the phone while I was in labor that because the labor was so drawn out, the baby would likely be born with spina bifida. No, I will wait until I see the doctors, until I see Simon, before I come to any conclusions. But by now, it is unconscionable not to have called Marc and Lorna, so I again borrow Dave's cell phone. The answering machine picks up and, after the sound of the beep, I fill in a few key details—fall, Air Evac, Vancouver General Hospital—but omit the more terrifying ones: the fixed pupil, the lack of feeling in the legs, the bleed in the brain. They had to operate, I say, but he's better now. He's improving. I tell them I will call when I know more, and end my message with the thought that is paramount in my head.

"Oh man," I say, "Simon will be so upset if he misses the gig this weekend."

This is the worst possible outcome I allow myself to contemplate. When Lorna hears my message, she is aghast. She wonders if I am in my right mind. Miss a gig? *That's* what I'm worried about?

But in the context of our daily life, this is not an idle concern. If Simon were to miss the gig, he would be disappointed, of course, but it would cut deeper than that. He would be angry, I imagine, and resentful. Not angry with me, but with everything. Our life. And so I continue to compose my loose and silent plea to the universe: Please don't let Simon miss the gig this weekend. Please don't let him miss the gig. Please. As far as prayers go, it is poorly constructed, narrow and self-serving, and so blatantly, colossally off point, but in my heart it encompasses a great deal.

IT'S BEEN A *long winter in my head.*

That's what Simon said to his sister last Sunday. She and her family were visiting Simon's parents in their farmhouse outside Montreal. Everyone had spent the weekend at the annual summer fair, and Simon was homesick. When he got off the phone, we sat in the front yard in the dying light of a warm summer evening. It had been a long, difficult winter for both of us, yes, but that day had been a good one. The three of us had gone out for brunch, and then we had dropped Eli at the floatplane that would take him to soccer camp on Vancouver Island. Back at home, we washed my car and his truck, a fun, soapy, not too taxing sunny-day chore. After, I finished reading the last few chapters of Dickens's *Great Expectations*—it was to be a summer of revisiting the classics—and Simon worked on his slide version of "Diamonds on the Soles of Her Shoes." A good day. Still.

It's been a long winter in my head.

It had been a particularly hard winter. My mother, who now lived in Powell River, a ferry ride away in the opposite direction of Vancouver, had been diagnosed with breast and lymphatic cancer. Over the past year I had spent a great deal of time in Powell River, helping out as she recovered from her mastectomy and began aggressive chemotherapy treatment. At first, this time was healing for our relationship, but as she became more ill, I found myself returned to the role I had inhabited as a teenager: the daughter who could never, ever do enough to help her mother. The daughter who was selfish through and through for wanting a life of her own.

"I love your mother, but she is driving me crazy," Simon railed. "She expects too much. It's not fair to you."

Simon was protective of me, and I appreciated that, but his anger didn't help. My mother had a potentially life-threatening illness. It was important to make, and keep, things right between us. Also, I suspected his motives. When he said that what she expected wasn't fair to me, he was also saying it wasn't fair to *him*. And he had a point.

He was tired. It's a difficult working life being a musician. The time dedicated to developing one's craft both as a player and as a performer is generally not reimbursed. As hard as I worked at my various writing projects, Simon worked even harder. A typical Monday-to-Friday work week for him meant getting up at 6:30 to do forty-five minutes of sight reading and scales on the guitar and then, after a quick piece of toast and a cup of coffee, working a full day, eight to five, of construction finishing work: hanging fir doors, laying oak floors, putting up beveled cedar siding on million-dollar homes. After dinner, from seven to eleven, he would either be rehearsing, with a band or on his own, or recording at the studio. On Friday and Saturday nights he played gigs, which kept him out until one or two in morning, and twice a month he worked the Sunday-afternoon jam at the Garden Bay Pub. Saturdays during the day, Simon watched Eli play soccer, and Sunday evenings he crashed in front of the TV and caught up on sports news, usually falling asleep on the couch. Monday morning, he'd start all over again. It was a demanding schedule, and while Simon tackled it with a seemingly endless reserve of energy that I both admired and was envious of, he was growing a little weary. That winter, in addition to being the primary parent while I was in Powell River, he had also finished up work on a major recording project, a CD by his band, The Precious Littles, entitled *Sometimes You Win*. A few weeks before, he had received the final mixed and mastered copy, the CD artwork was completed, and the unofficial debut was set for July 26 at the Islands Folk Festival, the gig I prayed he wouldn't miss.

> Sometimes it's coming up roses
> Sometimes dandelion wine
> You take your chances and honey
> Just jump right in
> Sometimes you lose
> But sometimes you win

Joe Stanton was the songwriter who penned these lyrics. Simon was the electric guitar player and co-producer, along with StraitSound studio owner Ray Fulber. Joe, Simon, and Ray had worked with the rest of the Precious Littles for over two years to bring the project to fruition.

The stress of creating a CD was over. Now the stress of releasing and promoting was to begin. *Sometimes You Win*, we'd say, as if it were a promise. A lot rode on the CD's success for everyone in the band. Like Simon, the rest of the Precious Littles were journeyman musicians, highly competent craftsmen who, in order to afford life on the Sunshine Coast, had to work a day job to support their musical career. A little radio airplay, some national exposure, increased CD sales: "winning" didn't mean striking it rich—it meant being able to support yourself and your family modestly by playing music.

Simon had started working full-time at construction five years earlier, when I returned to school, commuting to UBC to earn my MFA in Creative Writing. Now, with my thesis work nearly finished, it was Simon's turn to focus on music. He was brainstorming how to promote the new album, and he had lined up two new projects producing full-length CDs of original music, work that was to start in August. We didn't anticipate that he could give up carpentry altogether, but it seemed reasonable to think that his current work ratio could be reversed: in September, 70 percent of his work time would be devoted to music, the other 30 to various carpentry projects. These were big changes. When the mixed and mastered version of *Sometimes You Win* arrived, it was almost more nerve-wracking than exciting.

Still... the completed CD? It was something to celebrate. And so we drove into town together, bought a bottle of wine, and walked to Claytons, the local grocery store, to buy takeout picnic food from the deli counter. We were midway down the junk food aisle discussing the specific ways, come September, Simon could phase out carpentry work when he stopped beside the shelves of Perrier and tonic water, put his hand on the cart, and leaned toward me, off balance.

"I have such a weird feeling," he said. "Destabilized. Like everything in my life is about to change. Totally change, like on a molecular level. Like my very atoms are shifting."

"Yeah?" I said. "That's a good thing, right?"

"It's weird," he repeated and shrugged. He wasn't prone to weird feelings, much less to belaboring them, and although it was obvious that the sensation was an uncomfortable one, it seemed inconceivable that a premonition of change could bode anything but good. Simon had worked so hard for so many years, and he was so skilled at what he did. This was his time.

VANCOUVER GENERAL HOSPITAL is monstrous, a labyrinth of parking lots and entranceways and long halls, designed, it appears, to keep me from finding Simon. The panic, held at bay for so long, burbles over as Ryan and I try to find a parking spot, then an entrance to the hospital, and finally where the hell we are supposed to go. We hurtle down various corridors, and for the first time, I am rendered incoherent with panic. I want to stop dead and call Simon's name until he comes to retrieve me and explain what the fuck is going on. Rounding a corner, we meet up with Lou and Dave, the same look of unsuppressed panic on their faces. Then we see Simon's uncle Jerry and his aunt Barb stepping out of an elevator.

Jer is a judge, recently retired but still possessing both the ease and weight of authority his title conferred, an authority that is easily matched, if not exceeded, by that of Barb, his wife of forty years. It is a relief to see them. They are parents, and I have never felt more like a child lost at the mall. It is also frightening: Lorna and Marc have called in the reserves. Jer and Barb's presence means something is really, truly, terribly wrong.

Together we all enter the elevator, which Barb and Jer have just exited in confusion. The door closes, but none of us thinks to push a button. We stand in the motionless elevator for a solid minute trying to construct a plan when someone—Jer? Dave?—finally remembers to choose a floor. We navigate the route to the emergency room, where, upon seeing a lineup at admissions, I move decisively to the back of the queue. This is something I can do: wait in line. As if my visit to the hospital is no more critical than checking on the status of a patient who, say, needed a few stitches or a tetanus shot.

Lou walks past me, past the red stripe that the queued line is not supposed to cross, and through the doors into the ER. A moment later he returns, takes my arm, and leads me back to a nurse and a social worker. I am the daughter of a social worker, and I know what her presence means: this is more than just serious. She explains that Simon is still in surgery. She says he is very sick. His condition deteriorated in-flight: when he arrived at the hospital, he had high blood pressure and a low heart rate and was unable to breathe on his own. His left pupil was blown. (*Blown.* This word catches me, traps me: *blown,* suggesting leaves or cherry blossoms, colored glass transformed into pretty, light-catching baubles, kites, or candles on a birthday cake. It is an inexplicable word in the context of Simon's eyes, which are a pale, jade green with a thin amber filigree around the iris and which, most notably, are, and have always been, steady and wide open.) After having a seizure, Simon's body contracted into the stiff contortions of a decerebrate posture. Before being taken into surgery, he was rated multiple times by the trauma doctor as having a Glasgow Coma Scale score of 3, 3 being the lowest possible score and indicating the lowest level of consciousness. The nurse continues to speak, but the words blur into a smudgy, gray tangle in my mind: Catastrophic. Devastating. Life-threatening.

"I'll take you to the surgical waiting room," she says. "You can wait there."

GUIDO IS SITTING in a chair facing the door of the waiting room. He sits with the kind of effortless posture instilled through years of practicing the Alexander technique of efficient body alignment, a posture that evokes both grace and vulnerability. He is our inner circle, our closest friend since the early days in Montreal, and, for both Simon and me, like a brother. He is Eli's godfather, part of our family. He stands and holds me and I allow myself to exhale a long-held breath.

"Okay?" he says.

"Okay." As I inhale, my breath rasps down my throat, strange and shallow. The air, in this closed room, under the fluorescent lights, is all wrong. It vibrates and hums with an eerie dissonance. I sit beside Guido, woozy with a nightmare lightness. The social worker has already spoken to him, and there is no other news. Simon is still in surgery. And so we wait, time trembling to a standstill, a still life. The seven of us seated, silent. No words, nothing to say.

And we wait.

{4}

STAT CRANIECTOMY

SIMON IS IN a decerebrate posture when he arrives at the hospital: arms and legs straight and rigid and his head and neck arching backward. The zombie posture. A decerebrate posture is associated with a very poor prognosis, indicating that massive brain damage extends from the higher portions of the brain into the brain stem. A decerebrate posture means that internal pressure has caused brain tissue to be moved or pressed away from its usual position inside the skull. In Simon's case, the pressure from the buildup of blood in his left hemisphere caused the top section of the brain stem to shift down and to the right—a transtentorial herniation—so that the brain stem's fragile tissue bulged out from the narrow confines of the tentorial notch, the triangular opening through which the brain stem extends. The preservation of the brain stem's integrity is critical to a body's survival, as the brain stem controls our most basic functions—breathing, heart rate, blood pressure. And Simon's brain stem has been compromised.

The Glasgow Coma Scale (GCS) was developed by Graham Teasdale and Bryan J. Jennett at the University of Glasgow in 1974 to objectively evaluate the degree to which a person is conscious or comatose. The scale measures three responses: eye opening response, verbal response, and motor response. Each receives a score from 1 to 5, for a maximum score of 15, which indicates a normal level of consciousness. The lowest score is 3, which means there is no response in any category. Anyone scoring 8 or under is considered to be comatose. The general rule is that the longer someone scores 8 or less, the worse the prognosis is for meaningful recovery. Within the first thirty-six minutes of Simon's arrival at the hospital, the Emergency department records his GCS five times. The first three times Simon scores a 3—no response on all accounts—but the fourth time, after he is intubated and during his CT scan, he scores a 5 in motor response, for an overall score of 7, meaning that although he is neither speaking nor opening his eyes, he is responding to localized pain. He is sedated then, and six minutes later returns to being completely unresponsive and is taken to the operating room for a STAT craniectomy, "STAT" indicating that immediate, urgent medical intervention is required.

Here, a neurosurgeon named Dr. Haw removes a large portion of the left side of Simon's skull. Dr. Haw then cuts into the dura, the first of the protective layers that surround the brain, and begins the delicate search through Simon's left temporal lobe for the ruptured artery that is pulsing blood into the precious neural tissue of his brain.

He describes this process in an excerpt from his postoperative report:

The dura was incised in the temporal region with #15 blade and immediately liquid hematoma came out under high pressure. Further bone was removed to carry out a temporal decompression. The dura was then further opened in a large curvilinear flap and the hematoma was removed fairly easily using suction and irrigation. The underlying brain appeared to be in good condition. However, there was active arterial bleeding arising from the temporal lobe inferiorly. There was an evident temporal contusion, which was removed using suction and irrigation. The source of the bleeding was controlled using bipolar cautery and ultimately this gave good hemostasis [the process by which bleeding is stopped, which is the first stage in the healing of a wound]. The patient received frozen plasma. The brain was now gently swelling out through the dural opening. The dura was expanded and the bone flap was left out and sent to the bone bank for storage. The soft tissues were closed in multiple layers using staples for the skin.

A craniectomy differs from a craniotomy in that the bone that is removed to perform the surgical procedure is not replaced after the procedure is complete, as it is after a craniotomy. The skull bone is stored either in a deep freeze or, in an attempt to minimize the risk of infection, in a patient's abdomen or thigh, to be replaced, hopefully, at a later time. Both procedures have their historical roots in trepanation, the ancient surgical practice of drilling holes in the skull. Prehistoric skulls with holes that vary in size from a few centimeters to half the skull have been found in several countries across Europe. Trepanation was also common in Peru, where more than ten thousand trepanned skulls have been found, some dating back to the first millennium before Christ. Although surgical tools used to perform these surgeries were rudimentary—say, a piece of sharp obsidian, bronze, or copper attached to a carved wooden handle—there is evidence that many individuals survived the procedure and even lived for several years after.

That trepanation was performed, and often, indicates that early humans had some notion of the brain's central role in the body's functioning. Before Aristotle argued that the heart, not the brain, was the primary organ of rational thought; before Galen, surgeon to Ancient Roman gladiators, dissected the nervous system of oxen and coined the word *autopsy*; before René Descartes stated "I think, therefore I am" and attempted to separate the machine of a material body from his immaterial and immortal soul—before all this, ancient man was opening the skull for an array of medical, spiritual, and mythical reasons. Evidence suggests that trepanation was performed in an attempt to treat a range of conditions: depressed skull fractures, headaches, convulsions, and possession by evil spirits. Trepanation may also have played, for some cultures, a fundamental role in important rituals, informing both superstition and belief. Those strong enough, or lucky enough, to survive the dangerous procedure may have been honored as being blessed with special power.

Emil Kocher, in 1901, and Harvey Cushing, in 1903, became the first medical men in modern times to describe the use of a craniectomy to relieve the buildup of pressure in the brain. Despite long historical roots, craniectomies remain controversial. Termed a "salvage procedure," it is not subject to specific guidelines or protocols that state exactly when or in what circumstances the procedure is appropriate, and most research does not support its routine use in adults. The clinical difficulty is in knowing who is and who is not a good candidate for a particular intervention; the ethical difficulty is in employing an intervention that might convert certain death into a lifetime of profound disability. The decision to perform a craniectomy remains an individualized one, depending on the patient and the surgeon, although the most general guidelines indicate that a preoperative score of lower than 8 on the

Glasgow Coma Scale and the presence of one or more fixed and dilated pupils significantly increase the risk of an unfavorable outcome—defined as severe disability, persistent vegetative state, or death.

In Simon's case, by the time the neurological team had cleaned up the large subdural bleed in his left hemisphere, tissue had begun to swell out through the opening of the dura, the relatively inelastic membrane that surrounds the brain. The decision to not replace the skull bone was straightforward, because to do so would have compromised this bruised tissue. There was, simply, no space left.

I AM STILL waiting as this surgical procedure unfolds. Surrounded by the serious faces of Guido, Jer, Barb, Lou, Dave, and Ryan, I can no longer deny that this is really happening. So I phone Simon's parents and my mother. I contact Eli's best friend's father, a pilot, and arrangements are made to fly Eli in from Vancouver Island. Everyone will arrive tomorrow morning. Tomorrow. In this large generic room with a closed door, too little oxygen, and a terrifying excess of telephones, it is a remote concept, tomorrow. Tomorrow, it seems to me, might never come. At least not in the way the entire rest of my life has taught me to expect it to.

And we wait.

{ 5 }

IF...

E ARE STILL waiting.

As we wait, time narrows to a crystalline point between two opposing tensions: the longer we wait, the more our anxiety grows; the longer we wait, the longer we know Simon is still alive. I count the hours. If he was on his way to surgery at 1:30 when the hospital called me at home, then he has been on the operating table for at least five hours. Six hours. Seven hours. Jer and Barb have left, with promises to return tomorrow, when a nurse comes to tell us that Simon is out of surgery and is in the recovery room. At 9:30, another nurse, who introduces herself as Toni, asks me to come with her. She leads me down a winding hallway, explaining the layout of the floor, how access routes change when the cafeteria, Sassafras, closes at night, details that are so bewildering to me in the moment that she might as well be reciting complicated algebraic formulas. She leads me through the sliding doors of the ICU and into a conference room and tells me, once again, to wait.

A voice is shouting Simon's name. "Simon! Simon! Wake up, Simon!" I walk to the door of the conference room. At the end of the long hallway is a glassed-in room where a group of doctors circle a bed. "SIMON!" There is the sharp snap of a handclap. I move toward the glass room. The doctors turn as I enter, and I get only the briefest glimpse of Simon—his lips, his beautiful lips, so swollen—and the tubes and machines that are everywhere before someone—Toni, I think—is pulling me out of the room and ushering me back to the conference area. This time the door closes behind her.

"SIMON IS VERY sick," Dr. Griesdale, the head Intensive Care doctor, explains when he sits down at the conference table with me. It is the second time I am hearing this phrase, and this time it strikes me as profoundly wrong. *Simon is very sick.* It is grossly inaccurate, I think—more a concession to my current decreased mental functioning than a true representation of the situation. Simon doesn't get sick, I want to say. His twenty-four-hour flus last an hour, two at the most. Simon is strong. Not sick. But the doctor, pale and serious, is still talking. "Simon is very sick" is only the icebreaker. I instruct myself to pay attention. It is important to pay attention. Focus on the details.

"Simon has a serious brain injury," Dr. Griesdale continues. "It is life-threatening, and it will get much, much worse before there is any chance of it getting better."

There are three stages of a brain injury, he explains. There is the initial injury—in Simon's case, the fall—followed by the tissue damage caused by bleeding, and, finally, the damage caused by the subsequent swelling and by potential chemical changes in the traumatized brain. When Simon fell, he hit and damaged the back left side of his head. Inside the skull, the brain catapulted forward so that both the left and right frontal lobes slammed into the bony wall of the inside of his

forehead. The initial impact caused several skull bones to fracture. An arterial bleed in his left temporal lobe caused blood to
swell between the tissue of his left hemisphere and the dura;
during surgery, this large hematoma, or blood clot, in the left
hemisphere was removed, a vent inserted into the brain to
drain excess cerebrospinal fluid or blood that might pool and
cause an increase of pressure, and a craniectomy performed.
Because brains, like lips, will swell.

"The skull is stored in a fridge," Dr. Griesdale says. "If he
survives, we'll replace it sometime in the future. But you need
to be prepared. The damage to his brain is global; it is a very
serious injury. We are going to keep him in a medically induced
coma until some of the swelling comes down, but you need to
understand, he might not survive the night. He might never
wake up. And if he does wake up, he will never exactly be himself again. He will most likely be severely disabled. He will
certainly have profound and long-lasting brain injuries. We
will be making treatment decisions based on the potential for
his quality of life. Also," he adds, "because of the area of the
head injury and the fact that he could not feel his legs, we are
worried about his spine, but at this point a possible spinal cord
injury is a completely secondary injury. We will do an MRI scan
as soon as he is stable." He takes a deep breath. He's not done.
"There is the possibility of a blood clot forming at the top of the
cervical spine."

I do not understand the exact significance of this, only that,
like everything else, it is not good. A news piece I recently
heard on the CBC floats into my mind: the greatest amount of
damage to the spinal cord happens during the immediate post-
accident swelling. The quicker the intervention, the greater the
chance for recovery. I imagine Simon on a ventilator without
the use of his arms or legs.

"The blood clot . . .," I say. "Simon wouldn't want, I mean, if there is a surgical option, he would want to take the risk. For his arms." I don't know what I am saying, and a look I identify as pity crosses the doctor's solemn face. I want to turn away from his firm and unsettling gaze but cannot.

"A blood clot at the top of the cervical spine will not be life-sustaining," he says. "Let's just make it through tonight. You need to be prepared."

A brief and terrible pause. I feel resistant, pig-headed, stubborn. As if he intuits how unwilling I am to understand, the doctor repeats his previous prediction: "This is going to get much worse before there is any chance of it getting better."

No, I want to say. I want to argue, debate, wrestle, defy this seemingly central point. Nothing can be worse than this. But what do I know? This is a new land. I know nothing.

"YOU CAN SEE him, but only briefly," Toni says as she leads me out of the conference room. "There are more tests, the MRI..." Her voice trails off as we arrive at the sliding glass door where I stood—what, twenty minutes, an hour, a lifetime ago?—the doctors shouting at Simon to wake up.

Simon. Oh God, Simon.

Wake up.

I cannot take it all in. The machines that breathe and measure and pump and drain. The mad array of tubes—thin, wide, variegated, colored—inserted into his mouth, his skull, his arms, his chest. The bloody yellowish-orange iodine stain that covers the left side of his head, the side where the skull has been removed. His hands. The wide strong fingers that taste, forever, of guitar strings, the chewed-down nails, rimmed slightly with the day's dirt. His hands, as familiar to me as my own, lying still. So still.

Simon. Oh, Simon.

Wake up, Simon. Please wake up.

"YOU CAN COME back later," Toni says, but I am immobile. She places her hands on my shoulders, draws me toward the door, and hands me a cup of ice water with a straw. She suggests I leave and get a hotel. Get some sleep.

But I don't want to go. I try to pull my thoughts together, the loose questions and images on the peripheries of my mind. I try to compose myself enough to speak to the kind nurse. What I need to do, what is critical, is to be close to Simon. If I sit still enough and quiet enough and wait, surely this whole situation will straighten itself out.

"I want to stay," I manage to say.

She nods and shows me the smaller, more intimate and empty ICU waiting room. "These chairs fold out into a cot." She demonstrates. "I can bring you some blankets."

She is about to leave when I call her back and ask her to return me to the main waiting room. I am certain that if I am left to navigate the single hallway on my own, I will become irrevocably, irretrievably lost.

Back in the main waiting room, I find that I can trust my voice only enough to announce that we are changing locations. Lou looks at me questioningly, and all I can do is shake my head. No. No, no, no. But once we are seated in the ICU waiting room, I have to make an attempt at explaining. It is impossible to repeat some of the things Dr. Griesdale said. I do not mention the risk of a high spinal cord injury; I do not mention the possible blood clot; I do not mention treatment decisions based on quality of life. I do not say it will get worse before it will get better. I say that the injury is serious and that they have removed a portion of his skull to allow the brain to swell. There is a good chance Simon will have a lasting brain injury. When Dave, Lou, and Ryan leave, I finish the thought to Guido.

"The skull will be replaced," I say, tears coming now. Tears that are like tight explosions, tears that hurt, tears with no relief. "He said the skull will be replaced if he survives. If."

GUIDO OFFERS TO stay the night, but I ask him to go home so that tomorrow at least one of us will be rested. At eleven o'clock I return to the ICU, to Simon's room. The MRI has been completed, but the overnight nurse tells me the results will not be available until tomorrow. I don't believe her, and I consider making a fuss. I consider making a fuss from far, far away. What would a fuss look like? I don't really know, and, more to the point, there is no fuss or fight in me right now. I sit beside Simon and try to orient myself to the various machines and blinking red numbers. Shortly after, and despite the lack of fuss, I am visited by two young doctors. The first reiterates what Dr. Griesdale said: the brain injury is life-threatening. A spinal cord injury is completely secondary. They can do nothing to address the spinal cord until the brain is more stable. And that might not be for another ten to fourteen days.

"Do you understand?" he asks.

"Yes," I say. Resistance has transitioned into numbness and despair, and this makes me compliant, willing to agree to almost anything.

He leaves, and shortly after, the second doctor arrives.

"Dr. King," he says, shaking my hand and then positioning a chair so that he can sit opposite me, so close that our knees almost touch. He again repeats the previous doctors' message: secondary injury; the brain needs to stabilize first before anything will be done. He goes on to outline the finding of the MRI. There is no evidence of a blood clot or high cervical fractures. However, several vertebrae in the lower spine are fractured, and there is a dislocation at the juncture of the thoracic and lumbar spine: T12/L1. The dislocation has severed the spinal cord completely. Although it is impossible to do a full assessment because Simon is comatose, the fact is that severed spinal cords do not heal. Simon will have no sensation or movement below his waist. So relieved am I to hear of the lack of a blood clot or high cervical fracture that I barely register this distressing diagnosis. It lands with a muffled thud deep in my consciousness with a strict injunction: new information to be dealt with later.

"How fast," Dr. King says, "how fast everything can change."

"Yes." Had I known, really known this before? This is the central point of *The Year of Magical Thinking,* the exact point of departure for Joan Didion's exploration of how a mind, her mind, attempts to catch up to that surreal and shattering moment.

I have such a weird feeling. Like everything in my life is about to change. Totally change, like on a molecular level. Like my very atoms are shifting.

I sit beside Simon and I swear I can feel, can almost see, his very atoms shifting.

I sit beside him as long as I can, but when I am too dizzy and sick to remain upright, I return to the ICU waiting room. A young guy and girl are there (brother and sister? cousins?), and it is evident they are the night shift of a long family vigil. The young man is kind and helps me to adjust the temperamental recliner into a horizontal plane. I lie in a state of suspended animation and try to breathe. Tears come again, but when they do it is less like crying and more like vomiting, heaving and violent and beyond my control. I don't want to wake the young man and woman, so I leave the darkened waiting room to huddle under the fluorescent lights of the nearby bathroom. Around three a.m., I return to Simon's glass room with his Cormac McCarthy novel, and I read out loud so that he will hear my voice and know that he isn't alone. I read carefully, wary. I know that McCarthy's simple, clear language can lead a reader into sudden scenes of extreme violence, and I wish I had brought a different book for Simon. I do not want to stumble upon such a scene—no massacres of people or animals, no scalpings or vicious beatings. I scan each paragraph before proceeding, my voice a soft, undistinguished monotone that occasionally fades out when a stray thought overtakes me.

A stray thought like *It's been a hard winter in my head*.

The phrase Simon spoke to his sister two days ago. At the very moment Simon uttered those words, I was opening up a link sent to us by Jay Johnson, the drummer for the Precious Littles. The link led me to a TED talk entitled *My Stroke of Insight*, by neuroanatomist Jill Bolte Taylor. I watched it alone while Simon spoke with various members of his family, and was profoundly moved. As I made a pot of peppermint tea, I signaled him to wind up his conversation.

"You've got to see this," I said when he hung up the phone. I restarted the talk and we drank cups of tea while Bolte Taylor retold her amazing story of the stroke she suffered at age thirty-eight and of her subsequent recovery. On the screen, she held up an actual human brain, wrinkled and runneled, and trailing it, like a long hairless tail, the spinal cord. She discussed how the left and right hemispheres differ in what they think about and how they think about it: the left hemisphere, linear and methodical and constantly abuzz with the work of structuring our perceptions into a continuum of past, present, and future, versus the right hemisphere and its kinesthetic intelligence rooted in the sensory and energetic perceptions of the present moment. Simon, like me, was profoundly moved.

Shock and lack of sleep fuel the sense that this recent memory is heavy with import, rich with the possibility of hidden insight into our current situation. How is it that two days before an artery ruptured in Simon's left temporal lobe, we together, transported by Jill's tale, imaginatively constructed the possible effects of blood flooding the left hemisphere of the brain? But this sense of significance, however tantalizingly close, remains elusive. As I sit beside the machines and my husband in this insular glass room, it dawns on me that meaning might not have a place here. Maybe Simon and I having watched Jill's TED talk two days ago means nothing. Maybe this room, where I sit now and where others have sat before me, is the perfect setting for an unavoidable and brutal collision course with the fact, hard as stone tile, of meaninglessness.

Around five a.m., the neuro doctors arrive, a whole posse of them, for morning rounds, and they ask me to wait outside Simon's room. After their huddled meeting around the bed, the doctor who is in charge approaches me. He smiles, shakes my hand, and explains that he was part of the team that operated on Simon's brain. He has seen Simon's brain. He has, at least for the moment, saved Simon's brain. Suddenly alert, I catalog a series of irrelevant details: his lack of a doctor's coat, his paisley cravat, and his worn leather Dockers. He is fair and young and pretty, and I think, distractedly, of Dorian Gray.

"Simon's family will be arriving in a few hours," I explain. "Will we be able to ask you some questions?"

"No." His face clouds over momentarily. He only assisted; Dr. Haw, the lead surgeon, will answer our questions. He smiles again, courteous but distant, and says, "The underlying tissue beneath the bleed looked good. There is room to hope."

Several questions wing through my mind as he moves away. Hope for what? Hope that Simon will live, however altered or damaged? Or hope, real hope, for life, for Simon, for his beautiful mind? I want to call out in the otherworldly quiet of the ICU, call him back, ask him to clarify, but I am afraid even to voice the question out loud, let alone hear the answer. Room to hope. It's enough right now, no questions asked, to get me through the rest of the long night.

The nurses' morning shift change is at 7:30, and they ask me to leave for an hour. I retrace my steps to the ground-floor hospital lobby. When I return with a cup of tea, the lights in the ICU waiting room are still dark and the young man and woman still asleep. In the doorway a middle-aged woman sways. She clutches for my arm as I pass and asks when she will be allowed into the ICU. She leans into me, her breath sour with vomit and alcohol; the heat of her grief is overwhelming.

"In an hour," I say, backing away. "After the nurses' shift change."

She hangs on to my arm still, leaning in: a drowning woman. It is her son. He was beaten behind some bar on the Eastside. He is in a coma. She came in from Vancouver Island late last night, but she didn't know where to go. Who to talk to. She has no money, so she slept in the bushes outside the hospital. Her son is in a coma, his face beaten so badly it is no longer a face. Her son. Her baby boy. What is she to do? She is incoherent, grieving, desperate. Toxic. I have no ability to filter out the rawness of her pain: her pain is my pain is a world of pain opening up beneath me.

"The social worker," I say, pointing to an office door down the hallway. "Maybe she can help?" It is all I can offer. I forcibly extract my arm and move away, knowing I cannot be anywhere near this woman. The strength of her pain will pull me under. I will drown. I retrace my steps back to Sassafras, the large cafeteria adjacent to the ICU wing and the area that will become a meeting place over the next eleven days. *Sassafras,* I write in the notebook I carry with me. *A long bank of windows, many tables filled with interns, nurses, and people that are not you.*

{ 6 }

WILD THYME

7:45 A.M., JULY 23, DAY 2

I AM THINKING HARD about time. I would like to rewind it. Lines of a T. S. Eliot poem land in my thoughts like a whistling bomb:

For most of us, there is only the unattended
Moment, the moment in and out of time,
The distraction fit, lost in a shaft of sunlight,
The wild thyme unseen, or the winter lightning
Or the waterfall, or music heard so deeply
That it is not heard at all, but you are the music
While the music lasts.

I am shattered with clarity. I will myself back in time, twenty-four hours earlier, to my own unattended moment: I lie tangled in bedsheets. Simon stands at the foot of the bed, scowling. The day is warm already. The birds were particularly loud earlier, a disjointed cacophony that started just after four a.m. and that had, with sunrise, gradually become more shrill, more manic. They gather in our cherry tree where, too high to reach, the last of the tree's abundant fruit has fermented. Birds, keening, trilling, squawking: drunk on cherry wine. It is an annual ritual, this mad song of the birds, one that announces that the fullness of summer has finally arrived. The heat is a balm to my sore back, and as I stretch awake, I am porous, content, gloriously limber, and free of pain. I can smell the dusty, sweaty, sweet cedar of Simon's work clothes. He rubs his wrist. It's sore.

"It hurts," Si says. "But it's my last week. Shit." He wavers, irritable and indecisive.

"Eli's not here," I say. "We could play hooky together."

"Yeah, but Dave will kill me if I don't come in today," he says, resolving our brief discussion. He pulls on his shirt and heads for the coffee machine. If his wrist is still sore, he will take Friday off and give it a full day of rest before the gig.

This is the moment I return to over and over again. It is so close I can smell it, touch it, taste it. It seems possible, if I think hard enough, to return to it. This moment, before Simon takes his cup of coffee and walks the dog up the lane into the forest; before he gets in his blue and white beater of a pickup truck with the creatively rigged ignition. Before he leaves this moment, this time around, I'll grab his wrist, the one that isn't sore, and pull him back into bed. I'll say, *Don't go. Don't leave the house today. Stay here with me.*

{ 7 }

WEIRDOS

I MET SIMON AT Orangeville District Secondary School on the first day of grade 10, September 1984. This event has taken on a fabled quality in family lore, but the long-lasting ramifications of that first meeting were not immediately apparent, at least not to me.

"That's a great shirt," Simon said by way of an introduction. He spoke across the width of the third-period history room. He was fourteen, a year younger than me, and his hair was cut in a goofy bowl shape. He was tall for his age, still in what his sister refers to as his Baby Huey stage, a boy grown suddenly to the size of a man. I glanced down at my new T-shirt; it was baggy, to cover up my general scrawniness, the white cotton printed with an 1950s Roy Lichtenstein comic book–style drawing on the front: *Wouldn't it be wonderful,* a man in a bowler hat was saying to a well made-up lady, *if the world was full of weirdos?* It seemed an important question at the time.

"Thanks," I mumbled, a tangle of pleasure and embarrassment at being noticed, typical fifteen-year-old girl.

"No, really," Simon said. "It's a great shirt." He wasn't shouting, but his voice was loud and resonant, and it cut through all the pre-class chatter. For one brief, mortifying moment, the entire class glanced my way. The blood in my cheeks came to a slow simmer. Who was this bozo, anyway? New guy, acting like he owned the place.

"Quiet, class," the history teacher, Ms. Sodonis, said as she entered the room, thankfully ending my moment of mortification. "Everyone take their seats."

From across the room Simon smiled and gave me a thumb's up. Weirdo, he mouthed and sat down at his desk.

Hmm. Simon was nice enough, I decided, but in a loud, crude, boy way. Prim and calculating—I was hoping to befriend his sister, Emily—I decided he was, as a boy, not to be overly encouraged but to be politely tolerated, an attitude that lasted maybe half a week, until the day Simon invited me and my best friend, Veronica, over for lunch. The Paradis living room, painted sky blue and bathed in light from a large bay window, was home to the largest collection of record albums I had ever seen. That day we sang and danced to the Beatles, the Rolling Stones, T. Rex, Lou Reed, Taj Mahal—an explosion of sound and movement nested between biology and geography class that was transformative.

ALTHOUGH IT IS not entirely true, it could be said that before I fell in love with Simon, I fell in love with his family: his very poised older sister, Emily; his successful parents, Marc and Lorna; their beautiful, Victorian-styled home. The yellow canary named Bird. Because of Charlie Parker, Simon explained, and I nodded as if I understood the reference, which I didn't. The Paradis family had moved to town over the summer. Emily, it was rumored, wrote poetry. Lorna was finishing up her law degree. Marc worked at the Royal Bank developing computer systems by day; at night he came home and played guitar. Marc and Lorna offered up a radical new vision of what parents could be: When Marc unpacked the groceries on a Saturday afternoon, he bumped hips with Lorna as she made herself a sandwich. They would play fight for a tea towel, teasing one another. They laughed together; they sang.

Over the next few years, I received a comprehensive musical education under Simon's tutelage. I first heard Dylan's "Visions of Johanna" in his living room, and Patti Smith's "Horses." We swayed with psychedelic surrender to Donovan's "Mellow Yellow," thrashed around to the Clash's "Rudie Can't Fail," broke our hearts to David Bowie's "Ziggy Stardust," and had our minds blown by Hendrix's "Manic Depression" and Miles Davis's "Kind of Blue." I bought an old and scratched *Learn How to Disco* album at a church sale, and Simon, Veronica and I would practice the step-ball-step-turn moves, laughing until our sides ached, racing back to school between the first and second end-of-lunch bells.

That year Veronica fell in love with a boy, an artist who made beautiful silk-screened T-shirts, and they began spending all their time together. Our cozy triangle lost a corner. After school, Simon and I went to the little Dutch deli alone and ordered Havarti sandwiches on rye with black coffee and, our favorite, strawberry tarts, which Simon, in a moment of inexplicable whimsy, renamed "friendship tarts." One day he told me that his friend Cole wanted to ask me out.

"He likes you," he said.

"He doesn't even know me," I said. "I've never even spoken to him." Still, I knew who he was. Everyone knew who he was. A punk—an artist, too. A fine-boned beauty who dressed like Sid Vicious and who, like Simon, played guitar. They had been talking about starting a band.

"Well, he wants to ask you on date. Are you interested?"

"Sure, I guess." I shrugged, attempting a nonchalance I didn't feel. Unlike Veronica, I had never had a boyfriend. Had never kissed anyone in the back fields behind the elementary school. Never been invited to the parties where kids played spin-the-bottle in someone's basement.

For our first date, Cole invited me over to his house after school. He lived in a farmhouse at the end of a long tree-lined laneway on the outskirts of town. We set up a large panel of wood in a back field and pinned balloons filled with paint across its surface, then threw darts until we, the panel of wood, the trees, and the grass were dripping in an apocalypse of bright color. Back at the farmhouse, as we were rinsing our paint-streaked hands and faces, Simon phoned, his voice loud enough that I could clearly hear the question he asked Cole.

"Have you kissed her?"

"Not yet." Cole smiled and looked at me. "But I'm hoping to."

Later that night Simon biked over to my house and took the stairs two at a time up to my bedroom. He was quieter than usual, and he avoided meeting my eyes.

"Look," he said. "I think I made a mistake." He was pacing around my room, which was still the room of a child, filled with books, a dollhouse, and stuffed animals, the gray-blue wallpaper patterned with small white flowers, the pink satin laces of my ballet pointe shoes hooked over a nail in the wall. Simon was gripping a stuffed baby horse, his large hands wrapped in a murderous twist around its neck. "I asked you if you wanted to go out with Cole when what I wanted to ask was whether you would go out with *me*."

"Oh. Well," I said, my lips still shimmering from that first-ever kiss. I didn't know what to say. In Veronica's absence, Simon had quickly become my closest friend. I knew him. I loved him. But Cole? He was as fantastic and foreign as the far side of the moon. "Look," I said, scrambling for the right words. "The thing is, as soon as you and I kiss, we are going to have to get married and have a baby and spend the rest of our lives together. I think we should see other people first."

"Yeah, sure," Simon said. "Don't worry about it."

We were as young as that, once.

WHEN I WAS sixteen, my mother moved our family to Toronto, and Cole, not keen on long distances, dumped me, thoroughly shattering my teenage heart. Simon grew his bowl cut out into a mane of long dark curls and, with the help of a daily ten-plus cups of coffee habit, lost his dimply baby fat. He started a band called CODA, which played a mix of Zeppelin, Stones, and Police tunes. Always well liked by a wide cross-section of the high school population—browners, jocks, stoners, metalheads, preppies, teachers—he befriended them all, and his popularity increased exponentially when he started playing live shows. In his previous hometown in rural Quebec, he had been picked on, bullied daily, and his new social status in Orangeville was a constant surprise and gift to him.

He started dating girls two, three, four years his senior, often calling me for relationship advice. Despite distance, his dating adventures, and mine, our friendship flourished. We talked on the phone almost daily, and every few weeks I would ride the bus to Orangeville, or he would come to Toronto. We liked to stride through the city streets framing shots for album covers and making up band names, album titles, silly songs. We kvetched about the fickleness of our respective love affairs and promised that if things hadn't settled down by the time our hair was gray, we would marry. We sing-shouted the lyrics to the Beatles' "When I'm Sixty-Four" to one another. I knew these particular Beatles lyrics because I'd had to sing them, memorized, for a grade 8 assembly. Simon knew the lyrics because . . . he knew lyrics. I hadn't yet been able to name a song he couldn't sing.

Simon was a good singer who, despite being a musical perfectionist, always valued intent and commitment over technique. There weren't many people I would trust my enthusiastic but squawking singing voice with, but I trusted Simon. Despite my off-kilter notes, our voices together created an inexplicable sense of rightness, difficult to articulate beyond a blood-deep intuition of internal harmony or the effortless balancing of opposites: together we were risky and comforting, thrilling and safe. Long before our first kiss, Simon and I were singing in preparation for the long, sweet haul of our marriage.

WE DIDN'T WAIT until we were sixty-four. When we were twenty-one, Simon tracked me down to a small clearing on the edge of a large forest in Halfmoon Bay, on British Columbia's Sunshine Coast, where he and his acoustic guitar proceeded to seduce me—some might say unfairly—with song. Taj Mahal, J. J. Cale, Muddy Waters, Paul Simon . . . Simon's knowledge of music was encyclopedic, and he could play almost anything he had ever heard. And in between singing the songs that other people had written, he played his own music, instrumental stories that skipped and strutted and cakewalked down a long, winding path, stories that were exploratory, curious, open-ended. He refused to commit them to the confines of any song form.

"What's that?" I'd say. "It's pretty."

"It's nothing." This was his inevitable answer. "Nothing. Just an idea I'm working on."

But it wasn't nothing. When Simon sat on the porch steps outside my cramped kitchen and played, it was as if he was giving voice to some secret, urgent, inarticulate, tangled part of my soul. I was helpless against that kind of power.

We danced around each other for almost two weeks, each of us uncertain how to proceed. Then, one Friday night, we went to see a band play at a local pub. Out on the dance floor we swayed, a little closer, then a little closer, circling each other, and I realized what our problem was. A simple kiss would never do to break through the buffer of friendship we had built over the years. I leaned into his chest and bit his shoulder.

Rising the next morning, I took a moment to gaze at Simon's sleepy form. We had woken together so often over the years, him on an air mattress at the end of my bed, or me on the couch in his parents' TV room, that I wondered what would be different this time. What had changed now that he was in my bed?

"Your legs are gorgeous," I said, giving him an appraising look. "I never noticed before."

On Simon's twenty-first birthday, his mother, Lorna, called Halfmoon Bay to give him the news that he had been accepted to the music program at Concordia University in Montreal. It was the best birthday present. After high school, Simon had taken time off to play in a band and work. He had learned to play guitar by ear, never learning to read music, a liability he believed radically limited his chances of being admitted to an academic music program. So, in a burst of determination, he had devoted three months to cramming in two-plus years of Conservatory music theory in preparation for the entrance exam.

Simon's occasional but mammoth-sized crises of school-related self-doubt baffled me. Quick-witted and able to make spontaneous connections between disparate bits of information that were intriguing and often hysterically funny, Simon was inventive and resourceful—ingenious. The way his mind worked delighted me. No one I knew came close to Simon's intellectual precociousness. I wasn't surprised that he had been accepted into his program of choice. A little heartsick that he might leave me behind when he returned to Quebec, yes, but not surprised.

We decided that I would drive Simon to Montreal in my new-old gold Ford van and that during the trip we would make some decisions about our relationship. As far as road trips went, this one was mostly a disaster: ill planned and underfunded. The alternator sparked out in Renton, Washington, and three days later the battery died. We blew a tire on a desert highway before landing in Salt Lake City, cash-strapped and stranded, while we waited for Simon's parents to deposit some money in his bank account. With our last few dollars we bought a jar of instant coffee, a loaf of bread, some peanut butter, and a small bottle of bourbon, and we parked the van on a dead-end country road outside of town. It wasn't until Simon unscrewed the lid on the Wild Turkey and the fiery whiskey-sting filled the air that the possibility of pregnancy occurred to me.

"No, thanks," I said and waved the opened bottle away. A small print warning underneath the image of the bald-headed, ruddy turkey on the front warned *Alcohol consumption may harm a fetus,* and I knew then, suddenly, completely: I was going to have a baby.

I spent the rest of the trip calculating and recalculating dates. The summer had been hectic, and my memory was blurry. We had been careful but maybe not careful enough. It wasn't until we crossed back into Canada at the Detroit-Windsor border that I found the nerve to tell Simon I needed to take a pregnancy test.

"Oh," he said. "Okay."

We drove on for another fifteen minutes, silent.

"What will happen…," I said finally. "What will happen if we have a baby?"

"If we have a baby," Simon said without hesitation, "it will be our best friend. We'll be fine."

It was exactly the right thing to say. But after we returned to Canada, things—big life things—swiftly became much more complicated. While I went to my mom's new place in Kitchener, Simon went on to Montreal and prepared to start his first year of university.

"You *say* we can figure this out together," I complained on the phone during one of our epic long-distance talks, "but what you *do* is a different matter. You're out partying and moving on with your life. I'm just sitting here . . . pregnant."

I didn't know what to do. I was young and lost with no idea of who I was, where I wanted to go, or what I wanted to do.

"I don't know what to do," Simon said.

"Well . . . neither do I."

"I want to not know what to do together," Simon said.

"So do I."

And so the gold van took one last trip before she retired, delivering me from Kitchener to the neighborhood known as Notre-Dame-de-Grâce, NDG, in downtown Montreal.

On October 30, 1991, at the stroke of midnight, we read out loud handwritten vows and celebrated by eating avocado, crushed almonds, and red pepper in a garlicky dressing. The news of our pregnancy had not been well received by our family or friends—we were, according to the world at large, too young, too unformed, too irresponsible, too stupid, and too poor to have a baby, all of which we acknowledged was, to some degree, true—and there were no other people we wanted to invite. And so we set a third plate at the table for our midnight feast and invited the spirit world, an impromptu gesture meant to include every ghostly emanation, from Halloween spooks to our unborn child. There was no minister or justice of the peace, there were no flesh-and-blood witnesses, and I didn't change my name, but none of that lessened the commitment we made to one another that night—young, stupid, and poor as we were—to entwine our lives and raise our child together.

On March 7, 1992, a month early, I went into labor. At the hospital, hooked to a fetal heart rate monitor, I was cramped and scared, unable to get up and move around, and the only source of distraction was watching the tiny blips of Eli's heart rate. In the middle of the night, I realized that every time Simon entered the room and spoke, Eli's heart would speed up, as if in anticipation of their upcoming meeting.

Twenty-four hours later I was rolled into a birthing room, the bloody world inside me rupturing. Then, the sound of church bells. They were ringing for Sunday service, and the air in the delivery room trembled with resonant song.

"Hello, Eli," Simon said and held out our new best friend— milky white and wet, full of new breath—for me to see.

SIMON AND I have traveled a great distance together since the day we met in history class. We moved from Ontario to British Columbia to Quebec and back to BC, to the place where we have made our home, in Halfmoon Bay, not far from the site of our original kiss. There have been times, the good times, when we have embraced the need for compromise and constant reinvention, and there have been times, the tougher times, when we have resisted, even fought against it. But, always, we made the decision to put in the work. There were, of course, tangible benefits for doing that kind of work over the years—greater stability, security, steady forward motion—but the real reward is not so easy to articulate, residing as it does in the wordless late-night blending of bodies that had, ultimately, less to do with sex and more with the synchronized switching of spoons that occurred even as we slept.

What, I often ask my writing students, are the Horses of your piece? It was a question asked of me by my writing teacher, Maureen Medved, at the University of British Columbia. The horses, she explained, pull the covered wagon. The pioneers get in their wagon and start their journey west. Everything changes—the landscape, the weather, the relationships in and out of the wagon. The horses alone remain the same. She tapped the thirty pages of my fatally flawed manuscript and said, I don't see your Horses.

Some of my writing students understand the concept of Horses immediately. Others struggle with it. Do you mean theme, they want to know, or literary symbolism? No, I tell them, although they are connected. It's less about where a story is going, I say, and more about how it gets there. What is it, I ask, that pulls us through?

I believe that a marriage, like a novel (or any successful work of art), requires Horses. Good, strong horses: proud, tall bays with long black manes, possessing speed and endurance. Horses strong enough to pull you both through the changing landscape of your life, the changing landscape of your body.

Our Horses? Eli, always Eli, our first and most important work. Then the compulsion to do, to make, to create things both useful and sustaining, a compulsion that Simon and I share and that has led us over the years to bake bread, build houses, write stories, and record CDs. And then the standby steeds in our stable: good food, good friends, good family. And music. There has always been music.

{ 8 }

SUCKERPUNCH

─────────

A T SIXTEEN, ELI is tall, athletic, and strong. He shares his father's wide green eyes and the extra-plump lower lip of his mouth, a feature common to many in the Paradis clan, but he is longer and leaner than Simon. It is hard to pinpoint when I started reaching up to hug him, but it is still a new enough sensation as to feel odd and not quite right.

A serious soccer player, Eli has been attending a soccer camp on Vancouver Island. When he arrives at Sassafras in the morning, he knows only that Simon was injured in an accident, and his face is stiff and strained with worry. He is too big. It would be better if I could pick him up and hold him on my lap while I tell him.

I explain as best as I can. I strive to keep my voice as normal as possible, as if words like *coma, brain bleed,* and *paralysis* are routinely part of our conversations. Eli is silent while I talk, his hands clenched, his gaze lowered.

"Do you want to go see him?" I ask.

"I don't know," he says. He stands with a riotous lurch and shifts his weight forward, both hands reaching for the back of a chair, which I think, for a moment, he is going to throw. He is frightened, and his fear is unexpectedly volatile; a barely contained fury sparks off him. He glances in my direction. "I don't know if I can."

"You can," I say.

Together we walk down the long hallway and into the glass ICU room. It is a fresh assault: the tubes and the machines and vents and drains and parched lips. Still hands. Swollen face, the swelling worse today. Simon is a fallen giant, and beside me Eli shatters into a thousand pieces, the way I did the night before. *The first lesson of motherhood,* my friend Rachel Rose wrote in her poem "Notes on Arrival and Departure," *pretend you are brave / until you are brave.* I try. I pretend. But it is not enough for either of us.

"Dad?" Eli says. He stands still in the middle of the room, afraid to move any closer. "Simon?" His face has turned a fluid shade of oyster gray and sweat beads over his lip. I wonder if he will throw up. He turns to me, his eyes wild. "I can't stay," he says. "I can't stay here." As we leave the ICU, I am blasted by the force of his pain. It threatens to pull me under. I don't know how to be there for him and still keep my head above water, still breathe.

Back at Sassafras, people have begun to arrive. Guido, Lou, Ryan, Dave. My mother, who looks healthier now than she did even a few weeks ago. She has gained some weight, and the unnatural red chemo flush is fading from her cheeks. But her breath quickens just crossing the floor of the cafeteria. Her hair and eyelashes have yet to grow back, and her bald head is wrapped in a rust-colored cotton bandana.

"Your Georgia O'Keeffe look," Simon said when she came to hear him play the Sunday jam at the Garden Bay Pub a few weeks earlier. "It's working for you. You look great, Kathryn."

She holds me now, too tight, too long. I am suffocating against her shoulder. I am relieved to see her but also, irrationally, angry. She shouldn't be here. A hospital is the last place she, with her weakened immune system, should be.

"You must be so frightened," she says as she reaches for and holds Eli, and he howls, an animal noise, part grief, part anger, while he pushes her away and grabs my hand.

"I can't stay here," he says again. "I need to get out of here."

It is inconceivable to me to leave the hospital, so Guido takes Eli to the hotel room Dave has booked for us and paid for. It is a hard realization that I can't make this better, or safe, for Eli. But I feel some relief too that he has left with Guido. There is no one I trust more to be with Eli than Guido—that is why he is Eli's godfather—and now I can focus on the moment when Simon's mother, father, and sister will arrive. The moment when I will tell them what I have just told Eli. The moment when I have to repeat the litany of dire outcomes Dr. Griesdale outlined. The moment when I have to tell them Simon's spinal cord is severed. I have repeated this new information to Guido, Lou, Ryan, and Dave, and that was hard enough. Lou gasped and doubled over, suckerpunch to the gut, when he heard about the spinal cord. Since the moment he arrived at our kitchen door, his every action and gesture has told me that we can navigate this crisis together. That everything will be okay. But as he crumbles with this new update, it is clear we are all now passengers in a car that has spun irrevocably out of control.

The thought of telling Simon's family this is unbearable. In the pauses between breaths I continue to clutch at the thought that this can't be happening. It is a thought thick with desperate, childish need, a need that, with neither logic nor faith to fuel it, magically, savagely, continues to persist: if they don't come and if I don't say it out loud, it won't really be happening.

SIMON WENT BACK to Toronto in mid-March to visit his family—his parents; his sister, Emily; her wife, Sarah; and their two children, Oscar and Alice—while Eli and I stayed at home. It's been over a year since I have seen any of them, but here they are—Marc, Lorna, and Emily—sharp-edged and white with worry, standing at the entrance to Sassafras. We huddle in a small circle, barely able to hug hello for the weight of words hanging between us. Although I have been doing nothing but waiting, their appearance seems sudden, out of context. Bewildering. I feel as if I am meeting them for the first time, the familiarity of their faces only a pleasant déjà vu, a fleeting dream of a sweet but distant previous life.

In family photos, Lorna is the person who gazes directly at the camera, head tilted back, chin jutting out in a defiant thrust, suggesting a personality who is unafraid, even embracing, of confrontation, a personality well suited to her successful career as a criminal defense lawyer; but that bristling, pugnacious energy is absent today. She is deflated, trembling, her hand tentative on Marc's upper arm. Marc, a lifelong athlete who, in his mid-sixties, still plays tennis and hockey a couple of times a week, is the guy in the family who gets things done, who keeps things running smoothly. While Lorna always seems to step forward into the fray, Marc's most characteristic gesture, especially when he is particularly intrigued or troubled, is to take a step back, folding his long arms over his chest. This is how he stands now, waiting.

"Let's go somewhere quieter," I say, stalling. I lead them down the long hallway to a small, semiprivate anteroom midway between Sassafras and the ICU. Marc and Lorna sit across from me. Two Monet-like prints hang on the mauve wall above their heads: a stream running through an amber autumn forest and a path winding through spring-green trees. Emily sits beside me, leaning in toward my shoulder. She wears her

Accordion Revolution shirt, and it evokes a flutter, a shadow
memory, of happy days in Toronto: streetcar rides in the snow,
Earth Day on the Islands, Kensington Market, after-hours at
the Dakota Tavern. She is bracing herself for the news I am
about to deliver, but she is also lending support so I can get
through saying it.

I am afraid.

They will want to know why. Why were we on the Sunshine
Coast? Why weren't we in a city where Simon could more reliably work as a musician? Why were we so far away from his
family home? For his family, Simon and I have never stopped
being the rash young couple tripping into parenthood before
we were fully prepared. Our move west was an integral part of
our growing up, but I wasn't sure that's the way the rest of the
family perceived it.

And I am guilty.

Guilty for not reading the signs properly.

*I have such a weird feeling. Like everything in my life is about
to change. Totally change, like on a molecular level. Like my very
atoms are shifting.*

I am guilty for not listening hard enough, for not really
hearing what Simon said. I am guilty for not dragging him back
to bed, insisting he ice his sore wrist and play hooky with me.

But when I am done there is no anger, no recrimination, no
pointed finger. There is pain and there are tears and there is
fear and there is love. Unexpectedly, I am a little stronger having them here.

I take them to the glass ICU room, and Marc, like Eli, is
almost physically unable to walk into the room. He collapses
onto the nurses' station, his anguish a physical force. He
steadies himself and makes it in on his second try.

The left side of Simon's face balloons out from the gauze
bandaging. Every time I enter this room, the swelling is worse.

The ICU nurse has contacted Dr. Haw to report that Simon's left pupil, while still reacting to light, has become sluggish.

"But it's hard to monitor it," the nurse explains. "The swelling is so bad I'm not really able to open his eyelid."

{ 9 }

BREATHE IN,
BREATHE OUT

———————

MANY PEOPLE ARRIVE from the Sunshine Coast. I appreciate their presence and the energy they cultivate in support
of Simon, but I am unable to repeat the news of his prognosis over and over again. I have told Eli and Marc, Lorna and
Emily, and I can do no more. My mind is rubbery and full of
empty spaces and I am unable even to hear the well-meaning
assertions that Simon is strong. That if anyone can make it,
Simon can. The things they say I believe to be true—Simon *is*
strong—but all that is beside the point right now. I nod my head
blankly when they speak and feel as if I am agreeing to an obvious and potentially dangerous lie.

"Moments in time," my friend Susann whispers in my ear as
I lean into her hug. "These are moments in time."

This thought is a gift that acts as a buoy, a kind of life-saving
floatation device, and I find myself repeating it over and over in
the coming days just to keep my head above water.

THERE IS NO new update on Simon's condition, and in the early evening Marc and Lorna retire to Jer and Barb's house in North Vancouver. Like me, Emily cannot conceive of leaving the hospital, and she is determined to spend the night with Simon. She promises to call if there is any change in his condition, allowing me to retreat to the hotel room and spend some time with Eli.

The hotel room is a block away from the VGH Emergency entrance. It is almost seven o'clock in the evening when I leave, and the heat and the traffic have died down. It is a recognizable summer evening, similar to so many before. Still, the world outside the hospital is changed, sliding by on a different temporal tract; everything is out of sync. Disjointed daylight, the sun's rays splinter the air, fractal and weird and gleaming but drained of any real warmth or sparkle. Outside, life has sped up, and I have slowed down. Guido gave me clear directions, but it is only when I enter the lobby of the Park Inn that I am convinced I will be able to navigate the distance from hospital to hotel room.

Eli and I lie on the king-size bed facing one another, giving each other foot rubs while we talk. He wants to leave Vancouver and return home to the coast. He doesn't think he can spend long days at the hospital, waiting. He could go home and get a job. Do something useful. I don't know what to say. I don't want him to leave. What if he leaves and Simon dies before he can return and say good-bye? I want him beside me. Is that selfish? I don't understand how he can leave.

"I can't *do* anything here," Eli says.

He is like his father. It would drive Simon crazy to sit around waiting. Simon would understand and support this need of Eli's to not sit idly by. So I make a few phone calls, and a plan is quickly put in place. Eli's soccer coach, John, and his wife, Colleen, will take both Eli and the dog, Paloma. Eli will travel back to the coast tomorrow and return to Vancouver every few days to visit. I ask if he has any questions.

"If Simon... If he, I mean, will he have to be in a—a *wheelchair*?"

I hear in the rasp of his voice that this is the worst possible thing he can imagine. And so I tell him, with an assurance I'm not sure I believe, that a wheelchair is not the worst thing.

"Think about it," I say. "Simon likes *watching* sports. He doesn't like *playing* sports. He plays guitar, and he can do that just fine in a wheelchair." As I speak the words, I become more confident. "And he has the kind of personal resources, the kind of strength, that, as long as he could play music, he could deal with a chair. Right?"

"Yeah," Eli says. A reluctant agreement.

"I know we're not religious," I say, "but somehow we have to find a way to pray for his head. And his hands. If he has his head and his hands, well, then a wheelchair? We can work with that."

"If he stays there, in that bed, for a long time," Eli asks, "will he get a bedsore?"

It is a question that robs me of all my breath, a question so intimate, so raw and visceral, that I am almost certain my heart will shatter in my chest. A question that belies my sense that perhaps Eli is too young still to get it. He gets it. He is afraid, but he is still able to reckon with this, the hardest of human truths: that the body, despite all our grand schemes, is so utterly fragile. Vulnerable beyond our own believing.

"A bedsore?" I say. "I don't know, love, I don't know."

ELI TURNS THE TV on low and we lie together. While the TV is on, I drift in and out of consciousness—more in a state of troubled exhaustion than of sleep—but as soon as he turns the light off I lie wide awake. Adrenaline and a pure, basic fear form a clenched fist in my gut. I tell myself to breathe. And I do—I breathe in and out, lying still in a parody of sleep, until about two o'clock, when I leave the hotel and return to the hospital. Emily has retired to the reclining chair in the ICU waiting room, and I take my place beside Simon, beginning a pattern: Emily and I split the night into two shifts; she is there from ten in the evening to two, and I am there from two to morning.

Each time I enter this room there is a transitional period when I have to reckon with all the mechanical sounds and flashing red lights and high-pitched beeps before I can reach Simon, the human body beneath all the machinery. My breath pattern alters to match the slightly faster pattern of the ventilator, the airy inhale, the *whoosh* of the exhale—both inhale and exhale too loud, too short, and too symmetrical, reminding me with every breath of this basic fact: *Simon can't breathe on his own.* The noise of the machine would remind him of Darth Vader. I smile a little. Simon recently heard a comedy skit on the radio, a comedian from Newfoundland who translated memorable movie scenes into his hometown dialect. The *Star Wars* scene cracked Simon up, and for weeks he reenacted it every chance he got.

"I's your daw, Luke," I whisper to Simon, breathing audibly with the ventilator, a mock Vader voice. "I's your daw." The machine keeps up its robotic breath, and Simon doesn't turn his head and smile at me, the way I keep half-expecting him to, but still I am comforted by this shared joke.

To the right of Simon's bed is a large computerized panel that flashes a series of numbers, which I watch compulsively for clues to Simon's internal world, trying to discover a pattern similar to the one I witnessed the night Eli was born. Does Simon's heart rate speed up when I speak to him? Does his pulse quicken when someone enters the room? But there are no patterns here; the mini strobe lights of numbers fluctuate erratically. The most important number represents the amount of pressure inside Simon's head, and every time this number significantly changes I am compelled to share it with the night nurse. She is gentle and considerate with me, but after the third or fourth time I call her into the room, she offers some advice.

"It can be really easy to get distracted by all the machines and the numbers, but you don't have to watch those. That's what I'm here for," she says, bringing me a blanket that has been magically warmed and wrapping it around my shoulders. "You just need to be here with your husband."

When I am able to distance myself from all the life-sustaining tubes and pumps and monitors and just be with Simon, I find this time, the late night bleeding into early morning, peaceful. It is quiet and I am alone with him. I have heard that people are more likely to die during the long hours of the night, although I don't know if statistically this is true. It doesn't matter. This is when I need to be with him. Here, I breathe with Simon and the ventilator, lending my respiratory support. Breathe in. Breathe out. Breathe in. Breathe out.

It is also easier, when I stop number-watching, to observe the intricacies of the neuro-intensive care Simon receives. The nurse carefully charts every aspect of his bodily functions: heart rate, blood flow, oxygen and glucose levels, body temperature. The pressure inside his brain, the intracranial pressure, or ICP. All of it, more than I am capable of being aware of, is closely monitored or altered by her ministrations. Every whisper of Simon's body is dependent on her care. He is a patient: a nameless, broken man with little chance of recovery, absolutely, utterly dependent on the nurses for his survival. Anonymity, I think, must help doctors and nurses do their jobs, but I recall the words of my friend Rachel Rose. She told me recently of a project she had embarked on, writing biographies for the elders at the Louis Brier Home and Hospital so that even though they may have lost their mobility or their speech, their continence or their cognition, the doctors and nurses and caregivers would be reminded of the things these people had accomplished during their lifetimes: becoming a physicist, surviving the Holocaust, bearing eight children. And so I compose an introduction to Simon and our family to tack up beside the note reading *zero Left Side bone flap* that is posted over Simon's bed.

My name is Kara and I'd like to introduce you to my husband, Simon. I met Simon when I was 15 in grade 10 history class. We married when Si was 21, and our son Eli was born a year later. Si celebrated his 38th birthday last month with sushi, pecan pie, good Scotch, and a game of poker. His pet name for me is Stan, and I call him Beau or Baloo. He is the great love of my life.

The most important thing to know about Simon is he is a musician. He has a Fine Arts degree from Concordia University, where he studied jazz guitar. His passion is roots-based music and he has two bands: Gut Bucket Thunder, which leans toward the rock side of roots, and the Precious Littles, which leans to the country. Si is a pivotal and much-loved figure in the Sunshine Coast music scene.

For his birthday this year I gave him a trombone, an instrument he has never played. Initial attempts can best be described as melodic, rhythmic fart noises which cause spontaneous and uncontrollable laughter for me and Eli and acute distress for our dog, Paloma.

Simon is very close to his parents, Marc and Lorna, who still live in the farmhouse he grew up in, in Quebec. Also, his sister, Emily, and her wife, Sarah, and their two kids, Oscar and Alice, who live in Toronto. Simon's most enduring weekend routine is to rise a half-hour earlier than the rest of the house, drink endless cups of coffee, and chat with his parents or Emily.

Eli, our son, is sixteen and an athlete, a soccer player. Satur-
day is soccer day, and every Saturday Si and I are on the sidelines
(with more coffee) cheering him on.

Si is an improviser; he has a sharp, quick wit and a mercurial
mind able to string widely diverging bits of info magpie-like into
a cohesive whole. A musical mind. A mathematical mind. He is
a gifted talker: loud, gregarious, engaging, generous, charming,
entertaining. He's the guy in the room who puts everyone at ease;
the guy who pulls out a punch line for almost any situation. Si
doesn't just make people smile, he makes them laugh out loud.

For me, and many others, he has always been someone who
is relentlessly healthy. Gentle-hearted but rock solid. A fighter.
Impatient always, with his own weakness or vulnerability.

Some loves in no particular order: aged cheddar by the chunk,
banana bread, watching hockey, Paradis clan parties, CBC's Ideas,
intelligence, skilled craftsmanship, Eli-in-motion, a soulful musi-
cal moment.

Dislikes: kale, humorlessness, affectation of any kind, lack of
conviction, New Age-y faux spiritualism, sentimentality, medioc-
rity, smelly perfumes, slow drivers, gangsta rap.

My most profound thanks to all the people at ICU who are taking
such good care of him. He is a beautiful, good man, cherished by
so many: me, Eli, his family and friends.

I DON'T KNOW how often or by whom my wordy missive is
read, and I do not think it has any impact on the overall excellent care he receives. But I notice two things: when I arrive at
two in the morning, the on-duty nurse takes time to give me a
detailed update or, if they are busy, a note addressed to me by
name is waiting. *(Kara, Si had a pretty good evening. His pressures*
(ICP) have been much better and I've given him minimal extra

meds. I have not had to paralyze him at all. I'll be back at 4:00 and we can speak then. T) Also, now, many of the nurses refer to him by name. *Okay, Si, I'm just going to moisten your lips.* Or: *Easy, Si, we're going to roll you on your side.* And with this simple gesture, the use of his everyday name, he is, even in the depths of a coma, more present—the essential Si-ness of him—in that sterile, foreign, digitalized, glass room.

{ 10 }

A HOWLING STORM

JULY 24, DAY 3

THE DOCTORS—ALL NEW, unfamiliar faces and most so young they must be interns or residents—arrive for rounds before sunrise and once again explain to me that there will be no operation on Simon's spine until his ICP is stabilized. Yes, I tell them, I understand. And I do, a little better than I did the night before. I am learning the meaning behind the acronyms: TBI is traumatic brain injury; an EVD is an external ventricular drain, and there is one inserted into the right side of Simon's frontal lobe to drain excess amounts of cerebrospinal fluid, or CSF, to prevent a condition called hydrocephalus, which is an accumulation of liquid in the brain. But the two most important acronyms right now are ICP and CPP. Because, a nurse has explained, there is only a limited amount of room inside the cranium, the vertebral canal, and the relatively inelastic dura mater, and when there is an increase in any of the brain matter due to, say, bleeding or swelling, the ICP, or intracranial pressure, rises. When the ICP rises too high, it lowers the CPP, the cerebral perfusion pressure, or blood flow through the brain.

No blood flow means no oxygen, and no oxygen means cell death. ICP is measured in millimeters of mercury, and a normal rate for a resting adult is 7 to 15 mmHg. A consistent ICP of over 20 mmHg requires medical intervention, and if it remains over 25 mmHg, the situation quickly becomes critical. I have watched Simon's ICP levels throughout the night. When they rise above 20, a nurse comes in and opens the EVD and drains out cerebrospinal fluid until the pressure comes down.

"Simon's family have arrived," I tell the doctors. "They have a lot questions. Will we be able to sit down with someone today?"

There is some hesitation. A few of the huddled group—very young, definitely interns—avoid meeting my gaze. I am told Dr. Haw will answer our questions, but I have yet to meet Dr. Haw, and before a firm Q&A time is established, the doctors dart away, remote and insulated in their cloud of critically important preoccupation. Next to the doctors' professional detachment, I am achingly raw, feverish with fear and grief and lack of sleep, and I feel defeated by my attempt to communicate with them. I watch as the sun rises, its early-morning heat noticeably raising the temperature in Simon's room. It has been not quite forty-eight hours since he stood at the foot of our bed, maniac birds chirping outside, and rubbed his sore paw, but it feels like a lifetime ago.

The hour of shift change for the nurses is 7:30 to 8:30, and I am required to leave. I buy tea and a muffin for Eli and take them back to the hotel, where he and I make a few phone calls to finalize arrangements for his return to the coast. At 8:30 I return to the ICU, where I am greeted by Mike, the nurse from the day before. It is a relief to see a familiar face.

"I've got news," Mike says, smiling. "They are scheduling Simon for spine surgery later this morning."

"Why?" I ask. I don't understand. Not two hours ago I was told that there was too great a risk that his ICP would rise dangerously high during surgery. Earlier, Dr. Griesdale had told me Simon's ICP levels would not be stable and would likely rise over the first seven days as his brain reached its maximum amount of swelling. A panic surges through me, similar to the moment when we arrived at the hospital and I didn't know how to find Simon. "The spine doctor told me the cord was severed; surgery isn't going to change his chances of walking, right?" I ask.

"That's right," Mike says.

"But it could risk his brain, right? If his ICP rises? Why would they do that?"

"Well, if his spine is mobilized, it will be helpful to us in taking care of him," Mike says.

"That's not a good enough reason!" I shout.

"We'll get more information for you," Mike says, "okay?" Since Simon's family arrived, Mike has been nothing but kind and considerate. While he wasn't necessarily expecting me to be happy about this news, I don't think he was expecting this distress. "I'll talk to a doctor."

"I want to talk to a doctor," I say. "I won't okay this surgery until a doctor has explained the reasoning to me and the rest of the family."

I leave to go find Emily and my mother and to contact Marc and Lorna at Jer and Barb's house. An hour later, we all return to the ICU. Mike tells us Simon's surgery has been canceled because of an incoming emergency but was rescheduled for tomorrow. Dr. Griesdale will speak with us and answer all our questions later in the afternoon. The rest of the family is now at the hospital and the prospect of imminent surgery has been averted, so I leave to see Eli off to the coast.

I return alone to the hotel room to have something to eat and take a nap. Both concepts—eating and sleeping—seem foreign and impossible. Instead, I shower and cry. What if they canceled Simon's surgery today because I made a fuss? What if the surgery today could have prevented some damage to his spinal cord? What if I had grabbed him by the hand as he deliberated at the foot of the bed and insisted he stay at home? What if I had never gone back to school and put us in the position of needing the extra income from construction? What if we had moved to Toronto instead of Halfmoon Bay? Or Halifax? Or Vancouver?

Over and over again I relive the moment when he stood at the foot of the bed. I can still feel the early-morning heat of our blankets. I am guilty for not dragging him back to bed. I am guilty for not properly reading the signs laid out before us: Simon's moment of destabilization in the grocery store? Change, he had said, on an atomic level. And he rarely considered taking a day off. Why that Tuesday? I am convinced that the universe supplied all the evidence needed to predict this event, and I am guilty because I could have—should have— done something to prevent it. And I didn't. I am consumed by imagining how, by simply reversing one of those decisions, I could alter this series of events. How far would I have to go back to ensure that what is happening now would be impossible? Would I return to British Columbia with Eli and leave Simon in Montreal all those many years ago? Would I give up my life with him to spare him this? *Yes,* I think, *yes, yes, yes.* This is a dangerous train of thought, one that pulls me apart, fiber by fiber. I am insubstantial as a tuft of carded wool. And I would make almost any deal with the universe if it meant I could undo the precise moment at which he fell.

Back at the hospital, I sit with Simon's friend Sully, who, even in the summer heat, wears a plaid flannel work shirt, the cuffs frayed, and baggy, grout-stained jeans. I covet the flannel shirt, in all its tobacco-scented, oversized glory, and want one of my own to wrap myself in. If I asked, Sully would take the shirt off and give it to me now; he is that sort of friend. A big bear of a man, Sully plays bass and sings in a sweet Nashville tenor, one that Simon greatly admires. Sully tells me he wants to do Simon proud by being the kind of person Simon would be at a time like this. It is a common sentiment among those who gather at Sassafras. In this type of crisis, the person you most want around is Simon: strong, resourceful, constructive, positive. The kind of person who can be emotionally open and available without being crippled by those emotions.

"I know," I say. "Damn Simon. Taking a nap when I need him most."

"That's the thing about Simon," Sully says. "He always puts everyone's needs before his own. He always thought first about you or Eli. Not himself, not his music."

It strikes me then that Sully thinks I am guilty too. He doesn't say it to hurt me, but I know Sully well enough, I know the life of a musician well enough, to know there is more than a hint of an accusation in this statement. Strangely, it does not increase my guilty feelings. For an instant, I see how ridiculous and self-centered my guilt is, how absurd. Simon is not, and has never been, a martyr. We made decisions *together*. Countless important life decisions and countless seemingly unimportant ones, like taking Tuesday off versus riding it out and reassessing come Friday. My guilt is useless and counterproductive. Indulgent, even. Perhaps I prefer to cultivate my guilt rather than confront my helplessness. Like Sully, I have to learn how to rise to this situation. Be as strong as Simon would be if our positions were reversed. Eat something. Get some sleep.

AT FOUR O'CLOCK my mother, Marc and Lorna, Emily, Mike the day nurse, and I meet with Dr. Griesdale in the ICU conference room. Dr. Griesdale's synopsis of this meeting includes the following statement:

They [the family] understand that Mr. Paradis may not survive
& that if he does, his functional neurologic outcome remains
uncertain. They asked many questions pertaining to progression
of neurologic recovery which I am unable to answer. They appear
to understand how very ill he is. I also explained the rationale for
early fixation of his spine, but they understand the window for
this may be closing. This will be answered tomorrow.

"Appear to understand" is the key phrase. Dr. Griesdale's explanation for the spinal surgery is essentially that there is, neurologically, a lull before the storm. It will be in Simon's best interest to stabilize him now. I remain uncertain, but the combined expertise of the Neurosurgical, Spinal, and Intensive Care teams is enough to convince me to agree to the surgery. Dr. Griesdale reiterates that the central issue is still whether Simon will survive. They want to proceed with the spine surgery although he might not survive it; even without the surgery, he might not survive the next hour, the next night, the next week. The outcome of his injury is still to be determined, Dr. Griesdale says. Yes, we collectively agree, and then ask a barrage of questions based on the premise that of course he will survive.

"How long might he be in hospital?" I ask.

"If he survives," Dr. Griesdale says, "a rough time frame might be three months in critical care, three months in the hospital, six months in rehab. Or, possibly, an indefinite amount of time in a long-term care facility."

"How long might he be in a coma?" Emily asks.

"If he survives ... it is hard to say. Depending on how stable he is, we will begin to lighten the sedation in anywhere from seven to fourteen days. It is important he wakes up slowly, but you need to understand he might not wake up on his own."

"Will he," Lorna asks, a barely contained desperation straining her voice, "will he be able to play music?"

"Again," Dr. Griesdale says, "right now it is still a question of whether he will survive—"

I can't listen to any more. Emily, who has worked as a medical transcriptionist to support her PhD studies, is taking notes. I can trust she won't miss anything important. I have to get out of the confined space of the conference room. I stand up abruptly, apologize, and leave. Sassafras is quiet after a busy day of visitors. Sully, alone, sits in one of the cafeteria's central tables. I crawl onto a chair and lean into his flannel shoulder as a howling storm catches up to me. I shake and sob and he holds on until the tidal wave of grief recedes enough for me to catch my breath.

ANIMAL SPIRITS

A NERVE CELL, OR neuron, is composed of three distinct parts: the cell body, the dendrites, and the axon. Neurons respond to a stimulus with a nerve impulse, an electrical signal that travels rapidly throughout the body. Dendrites receive nerve impulses from other neurons and are usually short and highly branched, like little trees extending from the cell body. The axon is a single, taillike nerve fiber that carries impulses to the dendrites of other neurons and can vary in length from less than a millimeter to over a meter long. The dendrites and cell bodies of neurons constitute the so-called gray matter of the brain and spinal cord, while the axons, insulated in a milky white lipid and protein covering called the myelin sheath, constitute the white matter.

The brain has four main structural components: the brain stem, the cerebellum, the diencephalon, and the cerebrum. The brain stem is continuous with the spinal cord, and it enters (and exits) the cranial cavity through the foramen magnum (Latin for "great hole"), the large opening at the base of the skull. Nestled behind the brain stem is the cerebellum, or "little brain," and immediately above the brain stem is the diencephalon, the mid- or through brain. Extending anteriorly from the base of the diencephalon and the brain stem is the cerebrum, the largest portion of the brain. Divided into left and right hemispheres, the cerebrum is internally connected by the corpus callosum, a broad band of white matter containing axons that communicate between the two hemispheres.

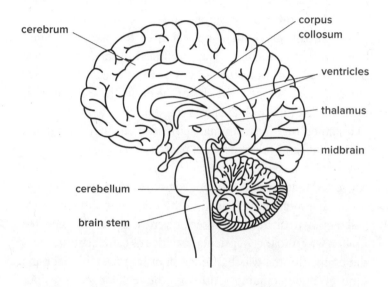

Inside the brain are four narrow cavities called ventricles: one in each hemisphere of the cerebrum, one in the midbrain, and one that lies between the brain stem and the cerebellum. It is in the ventricles that cerebrospinal fluid is produced and circulated, cushioning the brain's delicate tissues against the bony ridges of the cranial cavity and providing an optimal chemical environment for neurons to communicate in. The doctors have placed a vent, called an external ventricular drain, into Simon's ventricular system, by drilling a hole through the bone on the right side of his forehead. Through this vent the doctors and nurses are able to drain the pooling cerebrospinal fluid, relieving the buildup of pressure inside his skull. It was also here, in the ventricular system, that early physicians and philosophers named the locus for our humanity, the feature that most notably separates us from the beasts. They believed that the key to our uniquely human mental faculties lay not in the flesh and blood tissue of the brain but rather in the fluid tides of this inner space. For centuries, this was the dominant theory of the human brain throughout Europe and the Middle East.

Galen, the preeminent physician of the Roman Empire, believed that vital spirits were produced in the left ventricle of the heart and were carried upward in the carotid arteries. Once delivered to the ventricles of the brain, these spirits were transformed into animal spirits, the highest of spirits. Galen maintained that these animal spirits formed the "instrument of the soul" and, when needed, passed into hollow nerve cells to provide sensation or muscular movement. St. Augustine suggested that sensation belonged to the ventricles in the cerebrum, that memory belonged to the third or middle ventricle, and motion to the fourth, posterior, ventricle. Da Vinci injected molten wax into the ventricular cavities of cattle, cutting away the brain after the wax hardened in order to get a proper anatomical understanding. Although he challenged previous anatomical drawings with his work, he too unquestioningly accepted that the human faculties of imagination, memory, and intellect were located in the cavities of the ventricles. This basic belief—that animal spirits animate and elevate human thought by traversing the ventricular system—did not alter much until the eighteenth century.

Seated beside Simon, I am unaware of the historical importance of the ventricular system. Barely able to recall the basic macroanatomy of the brain and the microanatomy of a neuron that I gleaned in grade 12 biology, the brain and the infinite wonders of its gray and white matter are as mysterious to me as the dark matter in the deep space of our universe. Yet I need no history or biology lesson in neuroanatomy to understand this basic fact: as the cerebrospinal fluid is drained from Simon's ventricles, some vital life force is being lost. Here, now, the gradual dissolution of self from body that we all must eventually face is sped up to a frantic pace. Each time the drain is open, Simon leaks a little farther away from his broken body.

{ 12 }

YOU CAN'T
STAND UP ALONE

———————

MARC'S BROTHER JERRY and his wife, Barb, have opened their home, in North Vancouver, to Marc and Lorna, and tonight Emily makes the trip there for dinner. I read to Simon until she returns and takes my place beside his bed. I eat a bowl of soup and a piece of buttered bread at the tea shop across from the hotel and then lie down and sleep, finally, for almost four hours. I return to the ICU a little more grounded, a little stronger. Emily has drifted off to sleep in the chair but wakes when I arrive and we sit for a few moments, silent, both of us staring at the numbers flashing on the machine behind Simon's bed, staring as if we expect to see something more than a representation of an ascending or descending heart rate or spiking blood or intracranial pressure, staring as if we expect some deeper truth or future outcome to suddenly be made apparent. Emily sighs and stands up, a mild scent of rose hand lotion sweetening the antiseptic air. She kisses the top of my head and says, "I'll see you in the morning."

As I sit in the glass room at Simon's bedside, I recall the description Jill Bolte Taylor gave in her TED talk of the effects of the bleed in her left cerebral hemisphere, the ecstasy she felt, the simultaneous agony, and the difficult decision she made to fight to remain with her body. When I hold my hands on or near Simon's face, I can feel a kind of jumbled, roiling heat pulsing off his skin. I think I can actually sense the battle being waged at the peripheries of his physical self. *Please, please, please,* I have been praying, begging, pleading. *Please, please, please,* the supplication of a greedy and undisciplined child. But is my request fair? Would Simon want to stay, in this new, possibly profoundly damaged body? The thought that Simon might awaken from the coma to live a barely conscious life, seems, in this moment, more cruel than death. He would reject that. And so I revise my prayer and make him a new promise, which I scribble down in my notebook:

3:30 A.M., JULY 25, DAY 4
Hey, babe, I'm here. As usual, it's taking a little extra time for me to catch up with you. These last few days I've been locked in a back-and-forth struggle between absolute despair and blind hope. But no more. The reality is that this time is critical for you. Your injury is extremely severe but you are alive. And I am here with you as best as I possibly can. I'm here with you for as long as I possibly can.

Behind me, buoying me up, are all the many people who love you. It's huge. There is so much love and strength building around us, making sure that Eli is safe and I am able to be here, completely, with you. I am holding you strong, Beau, in my heart, my mind, my gut. Let me be your anchor. I know you might have to journey to a place where I cannot follow but I'm here, now. I promise I'll stay with you as long as I can.

AT FOUR IN the morning, Lorna calls the ICU. She is still on
Eastern time and is wide awake at Jer and Barb's hushed and
sleeping household. She is distraught. She and Marc remained
in the ICU after the family meeting and spoke with another
doctor, a young woman from Quebec, whose predictions for
Simon's outcome were even more pessimistic than Dr. Griesdale's. The young doctor explained that Simon's injuries were
global and diffuse, and if he survived, it was impossible to predict what areas might be affected. All of them, possibly. The
bleeding in his brain was extensive, and blood was toxic to
neurons; wherever there has been blood, neurons have died off.

"When I asked her about what the other doctor said to you,
that the underlying tissue looked good, she said, no. She said
his brain was soaked in blood. Blood soaked down to the ventricles." Lorna is weeping. "She said his brain was like soup
with neurons swimming in it."

Her distress is operatic, grief and regret pounding over her
in waves, words stuttering out, breathless and drowned. The
swells of her emotion make me feel absent, misplaced from my
own body. I am sitting at a desk adjacent to the nurses' station,
illuminated by the strident fluorescent glow of a desk lamp;
within its halo of brightness so lucent and heartsick, I am emptied of words. Finally, grasping, I say:

"You cut your hair. I've been meaning to say it looks great.
It's really beautiful."

Since the last time I saw Lorna, her long, wavy salt-and-
pepper hair has been shaped into a stylish silvery bob. "Simon
will love it." It is all I have to offer, and I hang up, leaving her
weeping at the other end of the line.

When I return to Simon's bedside, his ICP levels are spiking
up. Cerebrospinal fluid is drained from the external ventricular
drain, and now, instead of its usual clear color, it is pink with
blood.

I try to regain my earlier resolve. An image of a canoe comes to mind. It is a visual prompt given by my yoga instructor in certain difficult positions: Imagine that the back of your body is shaped like a canoe, your ribs the canoe's ribs, your inhale breath expanding and defining the shape of the boat. Exhale and relax your chest, your heart, into the boat's curved embrace. And so I imagine my body a canoe on a dark, cavernous river, wet walls of dark rock rising high on either side, eddies of water swirling in fast, confusing currents. Inside the body of the canoe is Simon, lying still to prevent further swelling, and in this way we float and bob toward morning, together.

DESPITE THE DIFFICULT night, the surgery on Simon's spinal column proceeds as scheduled, and we all reconvene at Sassafras. Guido brings a yoga mat, unrolls it in an uncluttered corner, and instructs me to lie down. His wife, Sari, and his cute-as-a-pumpkin daughter, Nadia, circle around us. Nadia, her round face framed by dark feathery curls, rolls pebbles off of cushions and crawls between chair legs, discovering discarded bits of food. At two years old, she is as willing to make a playground of a hospital cafeteria as she is Stanley Park.

"There's not much I can do," Guido says. "But a little Alexander technique? Maybe we can help your back make it through this."

Guido is a newly certified Alexander instructor, and over the last few years of his training, Simon and I have been willing guinea pigs. The exact mechanism of the technique is still mysterious to me, consisting mostly of Guido giving gentle physical cues designed to relax or realign areas where tension is chronically held in the body, but however it works, the results are undeniable: after a session with Guido I always feel taller and lighter inside my skin. An Alexander session is not like a massage; I am meant to be attentive to the process, a student of the technique not a recipient of a treatment, but now I lie passively, allowing Guido to readjust my shoulders and neck. My back is sore, and I'm sure Guido has noted how stiffly I'm moving.

Only a few months ago I suffered back spasms so bad that it became impossible to sit or sleep or move around, and I spent my days in a clutch of shooting pain. X-rays showed that I had a compressed disc at T12/L1, caused by undiagnosed fractures in those vertebrae.

"They are very old fractures," the doctor told me. "Can you think when you might have hurt it?"

I knew the exact moment. I had been thrown from an Appaloosa horse named Chocolate Chip mid-gallop across a field during a game of cops and robbers when I was eleven. After the fall I couldn't move. *My back is broken,* I had thought. *My mother is going to kill me.* The shock wore off seconds later and I got up, remounted, and continued our game, but whatever damage happened during that fall was finally catching up to me. The spasms were worse during the night, and on more than one occasion I was forced to wake Simon so that he could lift me out of bed, and I would hobble to the kitchen and lean against our fridge, reading, until the pain had eased. The spasms finally stopped, but the stiffness in those muscles still hasn't entirely gone away. It frightens me, this recent memory, saturated as it is with the absoluteness of my need for Simon. And now it is him with the crushed and dislocated discs, the severed, shattered nerves. I can't lift him out of bed. How will I ever be strong enough now that he needs me?

This sense of how much I need Simon strikes often and without mercy: I am strangled by a kind of mental barbed wire. I confess to Emily my crippling need, how over the years of raising Eli, my life narrowed so in focus. I haven't spent time cultivating friendships outside of our family. There was my writing work, there was Eli, and there was Simon: that was all I needed. Somehow I imagine life for Emily and Sarah in downtown Toronto is different; I imagine them surrounded by a vast network of supportive friends, but she assures me that, given the demands of work and kids, she too has reached a similar place of isolated and singular focus.

But despite my intense feelings of dependence on Simon and our relationship for meaning and connectivity, direct daily evidence of our community network—loving, resourceful, and generous—surrounds us. In such a short time, a very large group of people has mobilized. A network of rides brings Eli in and out of the hospital; salve is bought to apply to the anxiety hot spot the dog, Paloma, has developed; at our house the compost bin is emptied, the garbage taken out, and the lawn mowed. Patricia, Sully's partner, and Joe Stanton and his girlfriend, Sue, despite needing to ready themselves for the big weekend gig at the Islands Folk Festival, are making plans for a fundraiser. Dave has contacted WorkSafeBC, purchased a cell phone, and secured a hotel room for me. Every day, Simon's uncle Jer and aunt Barb take great pains to prepare a delicious evening meal for Marc and Lorna to return to after a long day at the hospital. In Toronto, where first-time home buyers Emily and Sarah were just about to move, a large group of people has arrived to clean, pack, and transport boxes. Emily connects via email to Simon's extended family and her own community of friends, many of whom have never met Simon, and receives, every day, long-distance messages of support and promises of prayer.

"You know we're here for the long haul," Guido says as he draws my left arm in a small, gentle circle in the shoulder socket. "There's a lot of people here right now, helping in the immediate crisis, but you know me and Sari and Nadia, we're here for the long haul."

I breathe into the sore spots of my back. It is possible, I think, that I can be stronger.

"I know it," I say. "Thank you."

SULLY AND MARC sit at a table and watch a video Sully has brought of the gig he and Simon played with their band Gut Bucket Thunder a few weeks ago. It was at a local elementary school, a charity event, featuring a catered dinner and a silent auction, and Guido joined the band as a special guest. I monitored Guido's stationary video camera throughout the night, adjusting the focus, pulling in for close-ups then panning out, readjusting the angle when the camera was knocked by the occasional elbow or hand of an enthusiastic dancer. I can't see the computer screen, but I can hear Simon's resonant voice singing the refrain of Jolie Holland's "Old Fashioned Morphine."

As much as I love Simon's visceral rendition of this song, hearing his voice through the thin computer speakers is unbearable. I want, so badly, to hear his voice, his real voice. I want to lean into the notch of his left shoulder, the one where my head fits perfectly, and feel the warmth and strength of his body. I want him—here—now—so desperately; it is unbearable to think that this Gut Bucket Thunder gig, so randomly captured on video, might be the final record of Simon in action. I leave Sassafras to go find Lorna and Emily, who have been visiting the hospital chapel.

Of all of us, Emily is the only one strong enough to continue to sing during this time. Her high, pure voice fills the chapel with an a cappella version of an old Martha Carson spiritual. *"You can't stand up,"* she sings, *"all by yourself. You can't stand up alone. You need the touch of a mighty hand. You can't stand up alone."*

EMILY, LORNA, AND I settle at a table on the outdoor patio at Sassafras, a safe distance from the video, which is still being played. As Lorna ends a call to her sister in Ontario, an urgent announcement comes over the hospital PA system for

the Greyson family to return to ICU immediately, and the
three of us reach for one another's hands, struck with the sudden, unwanted knowledge of what this type of emergency call
most likely means to that unknown family. Heart failure, seizure, a sudden turn for the worse—that is what an emergency
call to return to the ICU means. Are they gathering now, this
unknown family, at the bedside of someone they love while a
doctor turns off a ventilator machine and they say their final
good-byes? This is the moment, the emergency call, that we
have collectively been anticipating, dreading, defending
against all afternoon while Simon is in surgery.

But when the call comes for us, it is not urgent and the news
is good: Simon has been returned to the ICU in stable condition. At the moment, no doctor is available to talk to us, but the
surgery proceeded without complication and we are now able
to visit Simon.

We make our pilgrimage back to his room in groups of two.
Nothing has changed in the glass room, but there is general
agreement that Simon looks more comfortable now, that he
seems more settled in his body. We tell one another that the
occasional shuddering movement in his shoulders is a good
thing; it must be. He is fighting his way through the drugs and
the pain and unconsciousness back to us. He is struggling to
wake up. How could that not be a good thing?

This positive outlook is short-lived. When I return to the
hospital for the overnight hours, Simon's shoulder shudders
have transformed into rigors, a type of intense shivering that is
the body's response to fever. A cooling catheter is inserted into
his femoral artery at the top of his right leg, and it circulates
a cold saline solution through his body to keep his temperature down. I learn that Simon has not been able to tolerate
the nutritional supplement that is pumped into his stomach
through a tube in his nose; his digestive system has completely

shut down and will accept nothing. Nothing in, nothing out. And along with his bone-shaking shivers, he also has the hiccups, both of which contribute to a spiking ICP. It is over 20 mmHg, then over 25, then over 30. The external ventricular drain is open almost constantly, and the cerebrospinal fluid is distinctly colored with blood. Although the nurses continue to give him increasingly high dosages of a paralytic drug for the shivers and hiccups, every time I attempt to hold his hand or soothe his shoulder, he starts to shake. I feel that I am endangering him simply by touching his skin.

JULY 26, DAY 5

IT IS DR. King, the serious young doctor who spoke to me on the first night, who arrives at Sassafras to update the family on the spinal surgery. He explains that the surgery went well and that the spinal cord, despite being severely damaged, was not, as they had originally thought, completely severed. Nearly, but not completely.

"Although we can't assess him because of the coma state, we are assuming he is an ASIA A," he says. "Meaning," he continues in response to our baffled looks, "there is no sensation and no mobility below the level of injury. Which is from the waist down. But the injury to the spinal cord does not affect his ability to breathe on his own or to move his arms."

Even though that is essentially the same prognosis he gave me on the first night, I write in my journal to Simon that this is "the first real good news we have received." Good because the situation is not worse than we anticipated, and good because it introduces a margin of error into the doctor's predictions. This is the small margin—of error, of hope—that I want to reside in. Not knowing, Emily and I agree, is the most positive place we can realistically be. Like me, Emily is a writer, a researcher by nature, and for both of us there is safety to be found in the gathering and sorting of information. But an earlier attempt to Google some of the unfamiliar terms—*subdural hematoma, transtentorial herniation, decerebrate posture*—was quickly aborted. It is better, we agree, not to know.

MY BROTHER, HIS wife, Carol, and his twin three-year-old boys, Aidan and Lucas, arrive from Singapore.

"Oh, Sis," Rob says, holding me tight. "I'm so sorry."

They rent a hotel room downtown and join the growing contingent of folks who have commandeered a large corner of Sassafras, waiting. Waiting, waiting, waiting.

THERE IS A small pond beside our house; Paloma loves to splash in its muddy banks. At this time of year the mighty noise of the frogs fills the cathedral of the blue-black sky at night. As I walk down the ICU hallway, the thought of the frogs comes to me like a long-forgotten dream. The crickets are also noisy this time of year, and the birds too. Soon the cicadas will add their castanet-clacking to the summer's song. But here, summer has disappeared; only the hum of fluorescent lights and the changing of the nurses' shifts mark the passage of time. Time is static right up until the very moment when it races by in a kaleidoscope of nurses and doctors, in orthotic shoes and colored scrubs, pushing a stretcher, a mad, clattering migration to an operating room, winged with purposefulness. In their wake they leave a weighted hush, a palpable increase in gravity, a promise of winters that lie ahead.

As days pass, we are joined in the hallways by other families, who have come to wait out their own catastrophes. I see them leaning against walls, huddled in chairs, hanging on to one another. Although I am careful to avoid direct eye contact, with a brief glance I can predict exactly what stage they are at: they are in shock; they are in grief; they are, like us now, waiting. Waiting, waiting, waiting.

STAYING GROUNDED IN Simon's glass room becomes, in the wake of the spinal surgery, much more challenging. I don't need the nurses' updates—shivers and hiccups continue, the ICP is still rising, Simon is still not tolerating food, the fever is too high, glucose is too low—to know the situation has become more precarious. I can feel it. I can feel Simon, more distant and scrambled; I can feel him in the room almost entirely separate from his swollen body. It is as if the battle he wages has left the confines of his body and is being fought somewhere in the sterile air above and around him. Dr. Griesdale's clinical notes echo, in far less dramatic terms, exactly what I am feeling:

07/26/08
14:15 ICU STAFF
Will keep current management over next 24-48 hrs then consider lightening sedation but this is the critical period right now.

WHEN I LEAVE Simon's room, I am overwhelmed with anger, a very specific, direct, and irrational anger. I am not angry, as might be expected, at myself, for not earning a decent enough wage that Simon could just focus on his music. I am not angry at the culture of construction workers or even at the larger culture, for how we teach men in the crudest terms possible that caring for the safety of their bodies is somehow unmanly. (Hard hat? Harness? Pussy!) I am not angry that I didn't receive a phone call that might have got me to Simon faster, in his last few lucid moments. I am not even angry at the doctor who used the callously descriptive word soup to describe my lover's— my best friend's—brain. I am angry at the neurosurgeon with the paisley cravat, Dockers shoes, and youthful face who told me at the end of the first long night that there was room to hope. It is one of the kindest gestures a medical person has extended to me since Simon has arrived, and so the sense of its betrayal cuts deepest. I am furious.

The fabric of the day wears thin until I am clothed only in moth-eaten time. The unspooling of possible near-future events is terrible to contemplate, but what is worse are the past memories that descend with a clarity and closeness that is devastating:

Simon, returning home from work one evening when Eli is about seven. He is sad. On the road home, he saw a mama duck get clipped by a car, her ducklings scattering as she hobbled across the road. He jumped out of the car and tried to flag down an oncoming vehicle, a green Sundance full of teenagers, but they ignored him and hit the duck. Si said he could hear, could feel, the heavy death sound as she hit the asphalt. He searched for the ducklings that had dispersed into the bushes but couldn't find a single one.

Simon, at sixteen, calling to tell me he has tried pot and my prissy response that he shouldn't bother to meet me before the school dance, seeing as our lives are moving in such obviously different directions. Then the sense of relief three hours later when he arrives at the high school—the world has suddenly turned so boring without him—that I run across the gym and jump into his arms, a dance move we have perfected during long lunch hours at his house.

Simon, at twenty-one, two days after Eli is born. He has left to go do laundry at his aunt's house and, because of a faulty dryer and the distraction of an empty house in which he can practice guitar for his upcoming exams, is away for much longer than he intended. I am outraged at being left alone, a fatigued and nervous first-time mother. The following day, anger dissipated, I snap a picture of Simon and Eli: Simon, asleep on the yellow crow's-nest chair, Eli, glowworm-wrapped and snuggled so securely in the expanse of his father's broad chest.

Simon and Eli, only last Saturday, up at Connor Park, Simon helping Eli to perfect a set play penalty kick, the soccer ball curving neatly under the crossbar.

Waking, that first time, with Simon in my bed. "Your legs are gorgeous," I say. "I never noticed before."

I WAS SEVEN when my father died, and I have always attributed my sense of loss and aloneness to that early childhood experience of death. In *On Keeping a Notebook*, Joan Didion writes that "keepers of private notebooks are a different breed altogether, lonely and resistant re-arrangers of things, anxious malcontents, children afflicted apparently at birth with some presentiment of loss," a passage I recognized myself in when I read it in my early twenties. It is part of my process of magical thinking that I now see that all my earlier presentiments of loss were not, as I had previously thought, due to my father's death or a writerly quirk of my nature. Rather, the strange sense of loss I felt had always been leading toward this moment. Not as a preparation for it, but because of it. Simon's death, or his loss of meaningful consciousness, is a potential loss that is so big, I have, it seems, already spent my whole life grieving it.

Beau. Dear Beau. You know you are the music while the music lasts. Please let the music last. Please.

{ 13 }

SLEEPING BEAUTY

JULY 27, DAY 6

"LEGS, SHMEGS," I say to Emily as we lean against the nursing station, waiting outside Simon's room until the on-duty nurse finishes her routine assessment of his status. "Just give us his brain back."

"Oh, Stan." She reaches for my forearm. "I love you."

It has been a tough night. Emily lost her wedding ring, an event that would be terrible in any circumstance but that, in our current context, contributes dramatically to the sense that the entire world is unraveling around us. She correctly identified the hand-washing station at the doors of the ICU, where we stop upon every entry and exit to rigorously scrub fingers, palms, and wrists, as the last place of contact. As Emily scoured the floor and shelves around the sink, the sympathetic overnight nurse methodically uncrumpled every scrap of discarded paper towel from the emptied wastebasket and managed to locate Emily's ring in one such crumpled twist. It is back on her finger now, on the hand holding my forearm, a reminder that sometimes things work out all right. Sometimes, even when things seem irrevocably lost, they can still be found.

I feel lost. Lost in the seconds, minutes, hours that make up a single day. All I can ask of myself, all I can do, is hold fast to my promise that when I am in the glass room beside Simon, I will be strong and positive. I will be present. I will stay with him as long as it is possible. When I need to cry and whimper, when I need to remember and weep, I will go to the chapel or the hotel room. There I will allow my desperate need to surface, allow the subterranean prayer, the one that refuses to be silenced, the howling *please please please,* to find its full voice.

Simon comes to my mother in a dream and tells her to tell us to be patient, he is there. He is coming back. I distrust this dream and take no comfort from it, maybe because I receive no such dream-visits from Simon. I have only one dream. In it Simon is on the couch, awake and bored, and his legs do not work. I am rushing around, late for something, and Simon is trying to get my attention. I am irritated by the distraction and tell him so and it is precisely this irritation, this friction, that is comforting.

I want—am praying for—the luxury of being pissed off at Simon again.

"I keep thinking there is someone missing who needs to hear what has happened. Someone we haven't told yet. Someone who could make everything right," Lorna says as we stand in the cafeteria line at Sassafras, "and then I realize: It's Simon. He's the one that's missing. He's the only one who can make it right."

LATE IN THE evening, after Marc and Lorna have returned to North Vancouver, Emily and I arrange another meeting with yet another doctor, another new face, a young resident.

We ask questions about Simon's shivering, the fever, the glucose levels, and she does her best to answer. Although the meeting is at my request, it is Emily who does most of the talking.

I am saturated with information. And we have asked all these questions and heard all these responses before. I'm only looking for the margin of error in the doctors' diagnoses, and there is only one question I really need answered.

"What areas of his brain," Emily asks, "might be affected by the bleeding?"

"Well, the damage is global, so there could be a wide range of things: hearing, vision, movement, memory, language, personality. Intellect."

In other words: everything. I can't hear it all over again. "Can I...," I interject, finally screwing up the courage to ask my question. But it is almost unbearably difficult to voice it out loud. Once again Emily leans in to me, lending her support. "I need to know if... Is there... Is there any room to hope?" Up until the very moment I say this out loud and see the dismayed expression on the poor doctor's face, I realize I have believed she could answer this question, definitively. But, of course, she can't.

"Well...," she says, looking away then back. She smiles, ruefully but not unkindly, and meets my gaze. "... it's always a good thing to have some hope."

I AM WAITING for my bowl of soup at the tea shop across from the hotel when I notice for the first time a poem written on the wall in an ornate, cursive script.

> *Silently, silently I steal into my chambers.*
> *Deserted,*
> *Deserted and barren is the great hall.*
> *Waiting,*
> *Waiting for a man who will never return.*
> *Resigned I go to my tea*
> —*WANG WEI*

I cry as I read it. I cry most of the time I am not in Simon's room. For the first time, I retreat for a few hours to the hotel to make phone calls. I contact Veronica and some of our other mutual friends from high school. I cry some more. Then I shower and return to the hospital. It is only when I am in the glass room that I can find some strength and resolve. I let the rest of the world drop away so that I can reside in a place where death is no more mysterious than the taking or not taking of the next breath. Breathe in. Breathe out. I have made a promise to stay with Simon as long as I possibly can, and something inside is telling me we are coming to the end of our long, dark river. I don't dare touch his shivering, shuddering body, but I hold my hands an inch or two above his skin and imagine us floating along in our canoe-shaped embrace. It is a kind of a waking dream, this image, and right now it is all that exists.

By the end of the night, Simon's fever has broken: the ICP levels begin to stabilize, and the shivering subsides. Simon is tolerating the nutritional feed. The doctors initiate a new protocol to slowly lower his levels of sedation. I return to the hotel room but cannot sleep. Anticipation, aching and bright, is a punch of adrenaline straight to my heart. With lower levels of sedation will Simon wake up?

JULY 28, DAY 7
Nurses' note: 14:15 Pt. coughed and opened eyes slightly while being turned.

JULY 29, DAY 8
Nurses' note: 8:30 NEURO ASSESS: Pt. opens eyes to pain. No fixed stare. No purposeful movement. Abnormal withdrawal with both upper extremities.

SIMON FELL LAST Tuesday. He has survived the week. I start to write him daily updates in my notebook.

4:30 *a.m. I'm here, Beau. And you're here. We're here together.*
The doctors have decreased the morphine and sedatives you are
on. The last CT *scan showed there is still more swelling in the brain.*
It's okay, we were expecting that, but it should start getting bet-
ter soon, especially as you seem to be resting so peacefully right
now. You need to take a shit—you've been resisting that, I think,
with every molecule of your body—but of course it's critically
important.

Your body is breathing on its own a little. Every few breaths,
you out-breathe the ventilator machine. The tube in your mouth
is causing you distress but if you keep all this breathing up, hope-
fully they will remove it. Small steps, but they feel huge. One of the
nurses advised me to reflect on your progress, not hour by hour or
even day by day, but week by week. This seems wise. This time last
week, I was on the ferry desperate to reach you. I was being given
small bits of info and the potential seriousness of the accident was
washing over me, wave after wave of denial, anger, disbelief, horror,
and panic. You, Beau, were in surgery, doctors struggling to stop
the bleeding. You were, and have been, fighting to stay alive.

THROUGHOUT THE NIGHT the nurse continues to perform
neurologic assessments. She pinches Simon and yells his name.
"Squeeze my hand, Simon!" she yells. "*Simon! Simon! Squeeze*
my hand!" Each time she writes in his chart that although he
appears to be emerging from the sedation, he is demonstrating
no ability for meaningful contact.

JULY 30, DAY 9
3:00 *a.m. I'm here with you, Beau, and you're here, fighting so*
hard to wake up. Too hard. The doctors are concerned you are
spending too much energy resisting the sedation. You need to rally
all your energy to heal all the bruised and swollen places. You
were moving your shoulders and chest all night as if you were try-
ing mightily to wiggle out from under some heavy weight. You

*squeezed my hand, maybe. You heard me telling you over and over
again how strong you are. How loved you are.*

THE SENSE THAT Simon is fighting to wake up is so palpable,
I cannot leave the ICU, even though we are approaching the
7:30-to-8:30 shift-change hour, the time I am routinely asked
to clear out. I sit, clasping his right hand with both of mine, as
the nurses, preoccupied at their station, review charts. I lean in
and kiss his cheek.

"Hey there, Sleeping Beauty. It's time now. It's time to wake
up." I kiss the edge of his lips, still so swollen and parted around
the intubation tube. "Wakey wakey, rise and shine," I sing-
song. "You've had your sleep; I've had mine." It is the rhyme,
learned from his mother, that Simon always used to rouse
a sleeping Eli. I kiss him again and, staying close, cheek to
cheek, say in as steady and resonant a voice as I can muster: "If
you can hear me, Simon, squeeze my hand."

Squeeze. It is undeniable. *Squeeze, squeeze, squeeze.*

Communication! After a moment of this interaction, Simon
appears to drift back into his sedated sleep. I am able to update
both the departing overnight nurse and the arriving day nurse
before I leave to report this momentous breakthrough to
everyone.

It is a moment of exaltation. As Emily points out, this simple act is amazingly exciting because it means that Simon can
hear, that he can understand spoken language, and that he has
enough motor control to curl his fingers.

"All three useful skills for a guitarist to have," she says.

For an instant all the exhaustion lifts and is replaced by a
sense of absolute joy, a radiance that lights the world from the
inside out. But as the day wears on and exhaustion returns,
we are all shipwrecked on our small island of absolute joy
in a turbulent sea of fear. What we all want, of course, is for
Simon's eyes to open and for him to sit up and say *What. The.*

Fuck. Instead, there was a single hand squeeze and then a long, sleeping silence. When his eyes open, it is difficult to say what it is he is seeing. The questions that have tortured each of us—Will he wake up? And if he does, who will he be? What will remain? What will be lost?—are about to be answered. We have been told so often that he is unlikely to make a meaning-ful recovery that we are all scared to hope. So there is joy and there is fear. But also, and maybe most importantly for Emily and me, there is a new sense of purpose. Long after Sassafras has emptied of friends for the day and Marc and Lorna have returned to North Vancouver for the night, she and I remain, massaging and mobilizing the stiffness out of Si's hands.

We are updated by the night nurse, who again reminds us that if Simon really wakes up, his behavior might be unpredict-able and out of character. He might be violent. He might swear.

"Swear?" Emily says. "After what he's gone through? I'll be more surprised if he doesn't swear."

A doctor arrives in the room with consent forms. They want to perform a tracheostomy the following day, for several reasons. It is time to remove the intubation tubes: the longer Simon remains intubated, the greater his risk of infection, and the greater the risk of windpipe and vocal cord damage. It will be traumatic, the doctor tells me, for Simon to wake with the intubation tubes in his throat, yet it is critical to keep an airway secure. Simon is currently breathing along with the ventilator, but his body is not reliably regulating its oxygen levels. So they will make an incision in his trachea and insert a tracheostomy tube and he will be unhooked from the ventilator machine. I sign consent and Emily signs as witness.

TERRA, ONE OF the soccer moms, calls to tell me she is bringing Eli in to the hospital the next day and asks if there is anything I need from home.

"There's a block of unopened cheddar in the fridge," I say. "Could you bring that?"

JULY 31, DAY 10

Trach in today. After your surgery, you reached up and stroked my face. Also, reached down with my hand in yours and pressed on the top of your legs. I tell you that you have fallen and that you have hurt your head and your back and that we have been keeping you still to let the swelling subside. I tell you that this is a moment in time and that we are moving through it together. I tell you that you are strong, that your body is strong and that you can trust it. It knows exactly what to do.

Eli was here today and you squeezed his hand too. You'd be proud of him. He's scared but he's holding strong. Maybe too strong. He doesn't know how to feel sad. Not being sad, I think in his mind, is somehow a testament to your strength. He's angry, though. Your mom has taken some pictures of you and he would like to smash the camera. He'd like to punch the next nurse who pokes or prods you. He'd like to pull out the tubes and knock over the machines and pick you up and carry you to the ferry. Stomp his feet and say, Enough, now! Make this stop! Let's go home!

You expressed anxiety about the right side of your face, rubbing it. You said—squeezed—that it didn't hurt but you were worried. You said the trach hurts, but when I explained we needed to put it in to remove the intubation tubes, you raised my hand to your lips and kissed it.

It is a miracle to me, that kiss.

THIS IS A day of such great relief. Even though long stretches of silence remain, there is something so present in Simon's gestures. His right hand, eloquent and loving, reaches up to my face or rubs my back when I stretch in a way that tells him it is sore. I feel today that he is here with me, really here, really him.

Late in the afternoon I join our gathered friends in Sassafras. Hungry for the first time in over a week, I cut large slices of rugged cheddar—sweet and salty, smooth and granular—and pass it around.

LORNA'S EXPERIENCE OF the tracheostomy day is different from the one I share with Emily. Simon is being taken into surgery when she and Marc arrive in the morning, and she is horrified.

"Why is a trach necessary? He was taking long, slow breaths yesterday," she says. "I told him, 'If you take long, slow breaths, they will take the tubes out.' I told him that and he did. He was responding. He was taking long, slow breaths."

I outline what the trauma doctor explained the night before, but Lorna won't—can't—hear it.

"He was breathing on his own," she insists. "I would never have okayed it. They could give him an oxygen mask if he needed help breathing. Hospitals perform trachs the way they do C-sections. Most of the time it's not even necessary."

I too am horrified. I had thought that Lorna and Marc would be pleased with this progression. I wonder if part of her reaction is due to shock that Simon has returned to surgery yet again and her inability to control that event. Maybe she would feel differently if she had been involved in the decision. In hindsight, I probably should have called last night before signing consent, and consulted her and Marc. But then again, given her reaction, I am also relieved that the decision wasn't up for debate.

Later, when she sits with Simon, he touches the trach tube with his right hand, his face sad and questioning, and she feels as if she has betrayed the promise she made yesterday that the tubes would come out if he continued to breathe slowly and steadily. A nurse comes in to wash Simon's head and hair.

Lorna watches, horrified, as the nurse proceeds to scrub vigorously, moving Simon's still swollen head up and down. After unending days and nights of focused attention on keeping his head as still as possible, it is shocking. Traumatized, Lorna returns to Sassafras, where Emily, Marc, and I sit with Jay Johnson, the drummer for the Precious Littles.

"That bitch," she says, breathing fast and shaking visibly. "She was moving his head like this." Her hands jerk up and down. "He's on spine precaution! How can she not know he's on C-spine precaution?"

I don't know what to say. The ICU staff have been so exacting and attentive to every beat, breath, chemical shift, or electrical pulse in Simon's body that it is hard to believe a staff member would treat him so carelessly. There is only now, this very day, a small bloom of hope opening in my heart that maybe, with their careful ministrations, they have actually saved his life. Yet Lorna's rage is undeniable.

"I just want him to wake up and get all those tubes out. Have a cup of coffee," she continues. "Coffee would help. He always needs coffee to wake up."

Again, none of us know how to respond. Marc puts his arm around her shoulder and she leans into him. Jay tells us that over twenty-five years ago, after a serious accident on his bike, he was in that very same ICU and then was transferred to the spine ward.

"The hardest part," he says, "was having other people take care of your body."

"I don't think he even recognizes me," Lorna says. She pulls out of Marc's embrace and puts her hands to her face, weeping. "I don't think he even knows who I am."

Lorna's anguish reminds me once again of Joan Didion. When her daughter, Quintana, was recovering from a serious bleed in the brain, she too wanted to refuse the tracheostomy procedure. She writes:

In fact I had no idea why I so resisted the trach.

I think now that my resistance came from the same fund of superstition from which I had been drawing on since John died. If she did not have the trach she could be fine in the morning, ready to eat, talk, go home. If she did not have the trach we could be on a plane by the weekend. Even if they did not want her to fly, I could take her with me to the Beverly Wilshire, we could have our nails done, sit by the pool . . .

If she did not have the trach.

This was demented, but so was I.

I THINK LORNA is a little demented, but that is understandable. Being Simon's mother is different from being his wife. Simon and I wrote our own eclectic marriage vows, but, essentially, we echoed the standard marital contract: to love and care for one another in sickness and in health. This, then, is my chosen role, my primary focus: to love and care. As a mother, the central directive, however impossible, is to protect your child. And that is impossible now. Lorna cannot, no matter how much she fiercely wills it, protect Simon from the reality he is about to wake to.

THE GHOST IN THE MACHINE

AUGUST 1, DAY 11

SIMON'S RIGHT HAND talks to my body. He reaches for the spot behind my ear where for years he has drawn tiny half-moons late at night to help me fall asleep. He reaches for the exact cranky spot in my back, the place where he has smoothed Traumeel countless times into feverish muscle. He reaches even for the curve of my rump, demurely drawing away if a nurse enters the room. His fingers feather over the tops of his legs, tapping his hip bones and then, more insistently, over the top of his thighs, asking a question about the silence in his body that I cannot yet answer for him. But he is here with me, all of him, embodied in his active right hand.

THE PINEAL GLAND is small, reddish-brown, and pinecone-shaped and is attached to the roof of the third ventricle. It secretes melatonin, a hormone that helps regulate the body's biological clock by setting our sleep and wake cycles. It is this small, unassuming gland, centrally located in the ventricular system, that René Descartes claimed as the soul's conduit to the body, the dividing line in his mind–body dichotomy. Swaying in a sea of cerebrospinal fluid, the small movements of the pineal gland—according to Descartes—regulate the flow of animal spirits back and forth from the transcendent, immaterial world of the mind to the mechanical and mortal structure of the brain.

Although Descartes' theory of the pineal gland as gatekeeper between parallel states was largely discredited in his own time, his division of mind and body has exerted a long and troubling influence on Western society. For those of us, bookish by nature, whose inner fantastical world is vast and varied and who experience a sensation that our inner world runs concurrent to the day-to-day operations of the larger, outside, physical world, often with only occasional points of intersection, it is no great leap to imagine the impetus that propelled Descartes to differentiate between these two (often competing) realities. Still, as a young undergraduate, I, like my peers, rejected Descartes and his brand of dualism, with an ardor and forcefulness usually reserved for the villains and villainous acts of history. But now, sitting beside Simon, I find myself wondering if the mind, the so-called ghost in the machine, is truly a more ephemeral and transitory entity than I could have ever previously imagined. Where does the brain end and the mind begin? Is it possible, I wonder, that the mind, or a portion of the mind, can exist outside the brain's bowl of curds?

I ask this question because I believe that in my interactions with Simon's right hand, I am connecting with his mind. When I want to communicate with him, my first instinct is, predictably, to look into his eyes, but for the most part they remain closed. When they do open, they are unfocused and project only a mute panic: his right eye is swollen in a thin squint while his left eye bulges, neither working in concert with the other. It seems he is unable to move his mouth in the shape of words.

Perhaps he is also unable to find them or think them. But his right hand? His right hand knows what it's talking about.

WHEN EMILY SITS beside Simon at night, she sings. Most often, she sings Taj Mahal's "You're Gonna Need Somebody on Your Bond" and the Beatles' "Blackbird." Occasionally she sings an a cappella lullaby, "Golden Cradle," from an Emmylou Harris Christmas album Simon loved to hate, complaining outrageously anytime his sister played it over the holidays.

Tonight, her singing of "Blackbird" is interrupted when the beeps on the monitor indicate that Simon's heart rate is escalating. He starts pumping his right hand in a gesture she interprets as meaning terrible pain. She calls a nurse, and a dose of morphine is administered. As Simon's heart rate decreases, she tells him she is going to sing him "Golden Cradle" in the hopes that it will be soothing and will help him sleep. But if it bothers him, he should raise his hand and she will stop.

Before Emily can finish singing the opening line of this sweet, tender song, Simon's hand shoots up, the timing so perfect Emily knows he is poking fun at her and her love of Emmylou. It is the first time Simon's right hand has directly communicated with her, and, fittingly, it is a joke.

She laughs out loud. And doesn't sing any more Emmylou.

AT NIGHT, SITTING beside Simon, I often find myself thinking about Jill Bolte Taylor's TED talk, the one Simon and I watched just before his fall, the one where she held aloft a disembodied brain. In my memory the image of the brain takes on a distinctly animal-like quality: Jill might have been holding in her hands a rare and genius tree sloth or some variety of dynamic, long-tailed hedgehog, a much-loved but only partially domesticated and ultimately unknowable beast. I remind myself that her words, her story, illustrated an important point: the brain, that unknowable beast, with its origami folds and wrinkles, its faster-than-thought tracts and vast neural network, is resilient.

But as the child of a social worker and through my work as an advocate for adults with disabilities, I have plenty of first-hand knowledge that some brain injuries are profound and long-lasting. "Profound and long-lasting"—Dr. Griesdale's exact words. I know what that means. I have met, worked with, and cared for people with profound and long-lasting disabilities. In my life the closest I come to religion is my belief that a society is only good and strong insofar as it recognizes the worth, the inherent humanity, of its most vulnerable members. But facing this new reality, the possibility that Simon will wake up to a life with a profound and long-lasting brain injury, I am filled with a boundless horror and grief, emotions that are complicated by the snaking sense of shame I feel at experiencing them: Drooling Simon. Damaged Simon. Dumb Simon. Through all my years of advocacy work, it seems I have never really understood what is now a simple, basic, disgraceful fact: the significant reduction of a meaningful level of consciousness is a fate worse than death.

I push these thoughts away, over and over again, and return to the image of Jill Bolte Taylor's resilient and mysterious beast, and it acts as a kind of mental balm. I buy her book and, as I have done throughout my life when the problems of the world are too big to tackle, I start to read. As any unrepentant lifelong reader knows, good books are journeys, quests, vacations from the everyday, where, if we are lucky, we will discover as many interesting questions as answers. It is true: sitting beside Simon's bedside, I am oversaturated with questions right now, most of them too big and scary and amorphous to say out loud, but reading *My Stroke of Insight* helps—if not with finding answers, at least to give some of the questions a shape and form.

What is thought? Can it be located anatomically, in a specific structure of the brain or elsewhere in the body? Does it leave a physical trail in the tissues of the brain? Can thought be dissected? Deconstructed? Rebuilt? Can it be healed? Can Simon's thoughts heal? These aren't new questions; they have been around in one form or another since the Incan priests first sharpened obsidian flints in preparation for a trepanation ceremony. And the answers, throughout history, have depended greatly on where and when and who was asking.

During the Enlightenment, beginning roughly in the latter half of the seventeenth century and culminating at the end of the eighteenth as the French Revolution ebbed and flowed in the bloody streets of Paris, the twin values of information and communication reigned supreme. The publication of Newton's *Principia*, in 1687, marked a profound shift in thinking—the whole natural world, humans included, was subject to his laws of motion. Both telescope and microscope were invented, dramatically widening the depth and breadth of the universe. The popularization of the printing press was responsible for an unprecedented circulation of culture across all levels of society. Bach, Handel, Haydn, and Mozart became the chamber music rock 'n' roll stars of their day due to the patronage of a growing bourgeoisie who were willing to pay to have music moved from the courts of kings into their own domestic spaces. Denis Diderot's monumental *Encyclopédie* aimed to change common modes of thought by compiling all the world's knowledge in a single compendium that could be disseminated to the general public and future generations. Benjamin Franklin signed the U.S. Declaration of Independence and continued to indulge in the eccentric habit of flying a kite during lightning storms. Mary Shelley's Gothic monster, brought to life by a volt of electricity at the dawn of the nineteenth century, was the anxious bastard child of Dr. Frankenstein's Enlightenment ideas and ideals.

It was an age deeply skeptical of inherited privilege and divine authority, and the egalitarian and democratic sentiments of the philosophers, artists, and revolutionaries were reflected in anatomists' rejection of the idea that discrete functions can be located in specific, hierarchical structures within the brain. Instead, anatomists embraced the Doctrine of Equipotentiality, or the notion that there is an equal distribution of function throughout the brain. In this Enlightenment framework, the corpus callosum, the broad band of white matter that internally connects the two hemispheres of the cerebrum, took on a central role, replacing the ventricles or the pineal gland as the potential seat of the soul.

Giovanni Lancisi, an Italian papal physician, writes in 1712: "It is quite clear that the part formed by the weaving together of innumerable nerves is both unique and situated in the middle: and so it can be said it is like a common marketplace of the senses, in which the external impressions of the nerves meet. But we must not think of it merely as a storehouse for receiving the movements of the structure: we must locate in it the seat of the soul, which imagines, deliberates and judges."

More recently, in the age of the Internet and globalization—a modern epoch in which society is, as it was during the so-called Age of Reason, stitched together with threads of communication and information—there has a been a return to long out-of-fashion ideas about the brain and its potential, ideas that trace their lineage back to the Enlightenment notion of equipotentiality. With the introduction of the technology used in MRIs and other various advanced brain scans has come a revolution—not unlike the one that followed the invention of both telescope and microscope—in our understanding of the brain. In the last twenty-odd years we have come closer, from a biomechanical perspective, to discovering what thought is than in all the rest of human history.

As I sit at Simon's bedside through the long night, teetering on the waking side of consciousness, eyes puffy with the proximity of my dream-thoughts, a blue blanket warmed by the nurses draped over my shoulders, I ask these questions, not as a curious and disciplined scientist but as an overgrown child stunned by trauma into a state that hovers on the peripheries of everyday reality. It is magical thinking, a type of desperate whimsy, that questions where the seat of the soul resides: in the ventricles, the pineal gland, the corpus callosum, or an articulate and sonorous right hand? It is magical thinking that cultivates the intuition that the questions may be as important as the answers. That, perhaps, in some small degree, how the questions are framed and articulated will in turn shape and direct the answers. Despite their unscientific foundation, my conclusions and resolve are firm: The question is no longer "Can Simon's thoughts continue to heal?" The question is "How? How will Simon continue to heal?"

AUGUST 2, DAY 12

I SPEND ANOTHER long night in the ICU, leaving late in the evening for my midnight nap and returning early, unwilling to be away from Simon for too long. But unlike last night, tonight I worry that my presence is disruptive. When I am close, his right hand is busy, wanting to interact, but his communications are not as clear as they were. There is a desperation in his squeezes, as if he is shouting, *I'm in here! I'm in here!* to whatever question I might ask.

During the nurses' shift change, I return to the hotel and eat an egg with triangles of buttered toast in the quiet restaurant. I call my mother, who has returned to Powell River with my brother and his family, and give her an update. When I return to the ICU, a new nurse, one I haven't met yet, explains that Simon is to be transferred to the Step-Down Unit of the spinal cord ward.

"What? Why spinal cord?" I ask. "Wouldn't it make more sense for him to go to the ward for patients with head injuries?"

She looks at me as if I, too, am demented. "This is a good thing, you know."

None of us have spoken with a doctor since the trach was performed. All of Simon's vital acronyms—his ICP, his CPP—have stabilized in the forty-eight hours since the trach, but none of us know what to expect next. Dr. Griesdale is off the ICU rotation, and I don't even know whom to ask for. "Before he's moved, can I speak with a doctor?"

"I don't know. They don't come at my beck and call." She is clearly irritated. "This is a good thing. I don't know what you are so worried about."

Everything, I want to say, but don't. Right now there is fear and danger in everything. I retreat from her crotchety scowls to go find reinforcements. A short while later, I return with Emily, Marc, and Lorna, only to find that Simon is being prepared by another nurse and two aides for the transfer. This is happening. Simon is being wheeled out of the fluorescent womb of the ICU into the outside world, which has suddenly become a very dangerous place. How can Simon possibly be ready to leave the ICU? His fleeting moments of seeming lucidity are less, not more, stable than they were a few days ago. He cannot speak or move except for his determined right hand. He has no control over his bowel or bladder, his right eye is not opening, and his left hand is not moving. A wide tube in the hole in his neck continues to pump oxygen into his body. He isn't ready. None of us are ready.

SIMON'S NEW ROOM is private, extremely large, and immediately adjacent to the nursing station at the end of a long hallway in the spine ward. It is referred to as a Step-Down room, one where there is less focused attention by the nursing staff than in the ICU but more than in the regular ward. The room has a bank of windows looking out onto the Vancouver skyline, a private washroom, and, best of all, one of the big armchairs that unfolds into a cot. During the transfer, Simon develops a low-grade fever and a sharp crackle in his chest. At regular intervals, the nurses take a long suction tube, insert it into his trach hole, and vacuum out the phlegm from his lungs, an act of incredible violence—necessary and unintended violence, of course, but nonetheless violence that causes him obvious pain.

My good friend Mary, returned from a camping trip, is at the hospital along with our neighbor and friend Baelay. Baelay is as short and round as Mary is tall and broad, but they both have exceptionally strong and loving arms, and shoulders safe to lean on. They sit in Sassafras through the long day, simply there for whichever one of us needs them. Sully arrives too after his work day on the coast. He has been at the hospital almost daily, available around the clock for anything: a conversation, a drive to the ferry, a shopping list. In the evening, Marc and Lorna return to North Vancouver, Emily takes the evening shift with Simon, and Baelay, Mary, and Sully take me out for dinner at an Indian restaurant a few blocks from the hospital. When the waitress delivers my mango lassi, I burst into large hiccupping sobs.

"That's it," Mary says. "That's good. You have to let it out."

When I return to the hospital, Emily is sitting on the cot chair waiting for me. I am lighter without the weight of all those tears, but Emily is dark-eyed and hard-jawed, taut with anger and fear. She dozed off earlier, she explains, but woke just in time to see two nurses repositioning Simon on his left side.

"What are you doing?" Emily asked.

"He needs to be repositioned," one of the nurses began, launching into an explanation of pressure sores.

"No, no, no," Emily interjected, pointing to the sign posted above his bed. "He has no left-side bone flap."

Simon was hastily returned to his previous position. If he were to be rolled to his left side for even a short amount of time, his ICP could increase dangerously as the still-swollen left hemisphere comes into greater contact with the bony ridges of the skull opening.

"Keep one eye open," Emily says, kissing my cheek before she leaves.

One eye. Two eyes. I won't sleep, I know it.

Dangerous. Everything is dangerous.

{ 15 }

BEING WATER

AUGUST 3, DAY 13

First morning in your new room. I spent the night, as usual, but this time I actually had a chair/cot that I pulled beside your bed. Sunrise was beautiful—glittering Vancouver, pink-rimmed mountains, seagulls carving up the blue air—a real penthouse view. It was quiet and we lay together, side by side, for the first time in almost two weeks.

Earlier in the night, you were exploring your skull with your right hand. I explained about the bone flap. You rubbed behind my ear. You signaled for a glass of water. I didn't understand the gesture at first, until, when I was uncorking my bottle, you made the gesture with greater determination and I finally figured it out as I took a big gulp of water. I felt terrible not being able to give you a sip. I moistened your lips and explained that you are not able to swallow anything yet. But you are parched, a wild man in a desert. Your mouth and lips are so dry, and I can imagine how sore your throat is. There is likely a strep infection. You woke this morning with a fever and full-blown pneumonia, coughing huge chunks of lung goo into the trach tube. Oh babe, I want so much to make this all stop for you.

At lunchtime, after I left to shower at the hotel room, the nurses sat you up in that horrible bed-chair contraption. Your parents were told you were to be upright for about an hour, but your dad thought it was closer to two hours before you were laid back into bed.

You are not paying attention to your left side, and your right eye opens only a sliver. I feel you are more peaceful but less communicative. You squeeze my hand consistently, regardless of what I say. When I returned to the hospital and you were up in the bed-chair, I saw your gaze scanning the room, but I didn't feel like you were really seeing or hearing me.

IT IS IMPORTANT to get Simon up into the bed-chair as soon and for as long as possible to prevent the formation of blood clots. I know this, but still I hate the contraption. The nurses dress Simon in a blue gown and white paper shorts and white stockings called TEDs meant to increase blood circulation. They strap him in and tilt him upright. Here, his head and torso list to the left and he recedes to a distant, unreachable place. His mouth hangs open and his eyes half close. His beard has grown in thick and suddenly full of gray, and the remaining hair on his half-shaved head is long and straggly. Here, he is most un-Simon; with the swelling and listlessness and unfocused gaze and the crazy beard and redneck hair, he bears a frightening resemblance to the pictures of serial murderer Robert Pickton in the recent news.

Marc and Lorna remain throughout the day. Lorna sings Simon the story of his life since the moment of his conception, events unfolding in an invented and rambling choir-girl melody. Marc sings too. Sitting beside Simon, he holds his hand and sings softly in his ear. He sings the catalog of Paradis songs, the ones that are drunkenly hollered at large family gatherings and the ones that he and Simon played together sitting in their sky-blue living room on Sunday mornings. Bob Dylan, the Band, J. J. Cale, Little Feat. Music Simon has previously described as his uterine soundtrack or, more simply, Church.

I DISCOVER A quiet bookend of time at sunset and sunrise in the busyness of Step-Down. I write in my journal then, writing directly to Simon, an imitation of the whispered conversations that we might have back at home snuggled up in bed: *Eli is having a rough time. My mother is driving me crazy. Where will I get money for everything we have to do? I'm worried about you.*

These are the times that Simon is most present too. These are the times I am most clearly in contact with his consciousness, expressed entirely by his right hand. These are the times I *know* he is there. It has become more difficult to sustain that knowledge through the days. Every time Simon slips into a semi-lucid state, it is as if he is waking up for the first time; there seems to be no accumulation of consciousness from one waking moment to the next. The nurses have perfected a non-committal nod when I explain the "conversations" we have. In moments of doubt and despair, I wrestle with the fear that Simon is here with us but trapped within his mind, a bright spark adrift in a dark, internal sea. I wrestle with the fear that my need to speak with him is so great that I have imagined meaning in the movement of his right hand where none exists.

But these moments of doubt are brief. I can differentiate when his eyes are open and he is there from the times his eyes are open and he is not. Healing takes time, I tell him in those lucid moments. I tell myself the same thing: healing takes time.

Bit by bit I start to sing too. It is depressing to me that I have so little to offer; there are no fully realized songs on my tongue. My usually good memory fails me mid-verse, and a door closes on both lyric and melody. Nevertheless, I start to sing certain verses or even phrases on repeat: the middle section of Taj Mahal's "She Caught the Katy," Jeff Tweedy's version of the first verse of Woody Guthrie's "California Stars," Gram Parsons's "One Hundred Years from Now," the first song our family band, Farm Team (me on bass, Eli on drums, and Simon, of course, on guitar) learned to play together. I sing, but my voice is full of warbly pockets of air and uncertain notes. I worry Simon will be appalled at the quality of music he is forced to listen to.

"I'm sorry," I say. "I know it's terrible, Beau. I need your voice to lean on."

INSIDE ME, A deep well of panic forms. The farther away from the ICU we are and the longer Simon stays sleepy, the deeper that well gets. It is there all the time, and the only thing I can do when it threatens to open up beneath me is think of water. Cool blue water, surrounded by ferny banks and dappled with sunlight spilling through leaves and glinting off the mica in the rocks. I imagine diving into water; I imagine swimming in water; I imagine *being* water. Soft and yielding but containing an elemental strength.

It's a real place, the watery paradise I imagine: Scott's Falls. Simon and I biked there once when we were seventeen. We walked on smooth, round stones to the bottom of a large waterfall and splashed and swam in our underwear. It is a moment in my mind that existed so long ago and was so perfect, it feels more made up than real.

At night I wonder if Simon's soul will remain the same even if his brain has changed. I have never before thought about the soul as a distinct entity, because I never before saw the point of such speculations. But experiencing how thin the veil is between the parallel states of taking or not taking a next breath has created a new kind of openness in me, and I wonder if it is Simon's soul that speaks through his right hand. I wonder if that conversation is ours and ours alone and will go on, unchanging, regardless of what recovery happens within his banged-up body. These fanciful questions, however, remain forever unanswered and relegated to the nighttime hours; by day I work on gradually building a picture of the brain so that I can begin to understand what is happening inside him.

AUGUST 4, DAY 14

7:30 a.m. The nurses just kicked me out of your room so they can wash your back and do bowel care. There are bruises and horrible blisters on the top of your right foot from the boots ICU strapped on to prevent your feet from drooping, and a horrible sore on the back of your head, there maybe since the fall but big and mushy now from lying on it. Oh, the indignity of the body! I worry this will make you feel depressed. How could it not? It makes me feel depressed. Or maybe not depressed, exactly, just something different than the elation I felt when you first opened your eyes.

Early this morning, you reached your arm for me and put your hand on the small of my back. As if you know it's hurting me. It's impossible to know how aware you are in the semiconscious cloud you drift in and out of and yet, even in this state, you still are trying to take care of me.

The respiratory therapist was trying to get a baseline for your breathing capacity and, for the first time, you breathed more deeply when she asked you. This morning you seem less agitated and disoriented.

Your mom worries that you don't recognize her, although I am sure you do. I worry that your mother—in her frantic need to do something—is too much: too loud, too anxious, too stimulating. She is, like you, a person of action, and she feels there must be something that can be done to wake you up. She has been showing you tiny pictures of Ormstown on her camera's image viewer, images too small, I think, for you to focus on. She wants to put coffee in your feeding tube and give you a guitar. She blames everything right now on the tracheostomy, wishes that it had never been performed, as if this is the key factor in the speed of your recovery. I want to tell her that when I signed the consent for the trach, there wasn't a third, magical option for you to be all better right now. But I know she just wants to help. To do something. She, like all of us, is desperate, wanting so badly to lessen the pain of what your body is going through. And maybe she's right: maybe there is a certain type of stimulation that is the key to waking. Music, I think, must be the answer for you. The idea of bringing you your guitar right now seems beyond cruel to me but . . . you do need music. Bach, I think. And Bob Dylan.

Oh, Beau, these are tough days. I'm trying to be so solid in that room for you. To let you know that this is transitory, that you are moving through it (at your own pace), that we are moving through this together, that one day we will be on the other side. And I believe it almost all of the time. I am trying to know it, in every molecule of my being. But there are moments too when I feel myself surfacing from the shock; like you, I am waking from a dream-state, but still the nightmare continues. Is only beginning. For you: dealing with the indignity, the pain, the confusion, the changes in your body, the absolute lack of control. For me: figuring out how to be here, for you, while at the same time figuring out the new logistics of our life. For the glimpses of despair Eli and I see on your face. For all of us: the grief of being so brutally wrenched from the flow of our life. Walking in the woods with Paloma. Listening to you play your Dobro by the rock garden. Watching you promote the Precious Littles' CD. Picking blackberries and plums and getting wood ready for the winter.

Lao Tzu said, "When I let go of what I am, I become what I might be." I find myself repeating it like a mantra or a prayer when I stray toward too much sorrow or despair. There is

something there, in the Taoist's assertion of nonattachment, for all
three of us: a plea, maybe, to not hang on, kicking and screaming,
to the idea of our past life. Change is coming. I have found a kind
of solid peace by focusing on being present and in the moment, yet,
as I start to look ahead into our very uncertain future, it seems dif-
ficult, if not impossible, to sustain that type of presence in the daily,
larger operations of the world.

So . . . I am praying for my own strength as well as yours, for
courage beyond what either of us has yet to draw on. I am praying
for peace and resolve and patience and resourcefulness and adapt-
ability. I am praying, baby, for your mind. Your beautiful mind.

"WHAT IF HE is a vegetable?" Lorna asks upon arriving at the
hospital. We are standing outside Simon's room while a nurse
draws blood. "What if he ends up in a long-term care facility,
drooling in a corner, alone and sad?"

"That won't happen," I say. I am angry and exhausted by the
question, although Lorna is only articulating what we all fear
most. It is difficult for me to talk to her first thing in the morning. She arrives, recharged after a night at Jer and Barb's, with
the energy of a hurtling cannonball, and after a sleepless night,
I am often flattened by our exchanges.

Just as the nights are tiring to me, the days are long and difficult for her and Marc, and she is haunted by a story a friend
of Jer and Barb's told her. This woman's brother-in-law was in
a car accident and now lives in a group home because he never
made it into rehab. The rehab center had decided he wouldn't
benefit from therapy because he could no longer learn. For several years he cried daily for his wife to take him home. Once a
week they would take him by taxi to a physiotherapy class, but
no one expected that he would ever leave the care facility.

Lorna is also still haunted by the tracheostomy and the
image of the nurse vigorously washing Simon's hair. She is

convinced that the operation caused Simon to have a stroke and has been campaigning to have me, the nurses, and the doctors acknowledge that Simon's condition worsened immediately after the trach and the hair wash, arguing this as she might a court case. I don't agree that his condition worsened after the trach—in my experience, it briefly but dramatically improved—but in the last twenty-four hours, as I've felt him drift farther away, the constant campaigning has become more upsetting.

I leave the hospital and, after a nap, meet up with Guido and Eli. Guido has been searching for an apartment for Eli and me and has found one, three blocks from the hospital on Willow Street. It is a tiny attic space, clean and airy as an IKEA showroom, rentable by the month. There isn't much room for both Eli and me, and the rent is more than I can afford for any length of time, but I sign a damage deposit check. I will remain at the Park Inn for the rest of August, and now Eli and I have a place close to the hospital, at least for September. We'll figure out October, I tell myself, when the time comes.

When we return to the hospital, Simon reaches for Eli's hand. I am certain he smiles too. When Guido leaves the room, Simon moves his hand from side to side. A wave? Later, as Eli and I sit on either side of the bed, Simon, for the first time, squeezes my fingers with the up-until-now dormant left hand.

Marc and Lorna have good news to report too. Today, as Marc sat beside Simon, holding his hand and singing the Band's song "The Weight"—a family favorite—Simon squeezed the beat of the song in Marc's hand.

"It was at the 'and... aaand... aaaaand...' and he squeezed... squeezed... squeezed," Marc says. "He's really in there somewhere, isn't he?"

He's really in there somewhere. We all take turns saying it: Emily, Lorna, Marc, Eli, and I. He's really in there somewhere.

MARC AND LORNA take Eli to the ferry, and Emily leaves for a nap. We are switching shifts: I am doing the early part of the night, and she will do the long hours into the morning so that I can get a little more than two or three hours of consecutive sleep. She is leaving in two days, and I have already started missing her. It's hard to imagine not having her here, even harder, I think, for her to imagine leaving right now. I update my journal in the fading light of the sunset. Simon is asleep except for his constantly moving right hand. It's as if it is over-tired, overstimulated, and doesn't know how to shut down for the night.

"Hmmm," the new night nurse says after she performs a particularly brutal suctioning of Simon's lungs and takes his temperature. "He's still running a fever."

AUGUST 5, DAY 15
Doctor's History Sheet:
Pt. minimal response, zero verbal, blank stare.
Right upper extremity good movement.
Left upper extremity complete neglect.
Some confusion w commands, slow to follow but does.
Plan:
Repeat CT scan.

WHEN LORNA, MARC, and I arrive in the morning, Emily is upset: Simon's pneumonia does not seem to be responding to the IV antibiotics, and throughout the night, the gestures of his right hand became more vigorous and compulsive and less and less connected to any clear communication.

We sit down with the head nurse and request a trial with-drawal of all the morphine that Simon is on. It is impossible to say what Simon is experiencing in his body exactly, and while the last thing any of us want is to increase his pain, we all agree that we need to rule out morphine as a factor in his return to consciousness. The nurse, at first, is resistant.

"Extreme pain itself can interfere with a return to consciousness," she says.

"He has consistently told me that he isn't feeling any pain," I tell her. She is unconvinced, but we, as a family, are united in wanting this. Lorna's skill as a lawyer shines here, and she argues convincingly, refusing to take no for an answer. Marc contributes a story of his own bad reaction to morphine after surgery and a life-threatening infection years ago. The nurse agrees to consult with the doctors tomorrow morning.

AUGUST 6, DAY 16

EMILY IS LEAVING today.

It has been two weeks since she arrived, two weeks that encompass a lifetime—every moment, every breath in, every breath out accounted for. Together, we have slipped into a natural rhythm, ensuring a twenty-four-hour presence at Simon's bedside. She and I have shared stray recollections from our adolescence and conversations about relationships, aging, and trauma, but for the most part we have focused on the moment-by-moment task of making things work in this new, strange world, and we have accomplished this with blessed little need for conversation. We have shared the gift of being alone together. It is Emily's presence, so full of strength and song (and so like her brother's), that has kept me from being unbearably unmoored and lost during this time.

She has spent her final night with Simon, and when I arrive at the hospital in the morning, she is unusually rattled, even more so than the previous day. It has been a bad night. Simon's right hand was ceaseless in its "talking," becoming increasingly manic and jumbled, a compulsive tic-like movement rather than communication. After so many days and nights of constant vigil, Emily admits to feeling panicked at the thought of leaving, especially as it seems Si's condition is worsening. But back in Toronto, her partner, Sarah, and her kids, Oscar and Alice, are preparing to move from the family home, where Oscar and Alice were born, to their new house. Emily is needed there. And so as Marc packs her bag into the car, I promise to keep her regularly updated, to call whenever I need to, even if it is in the middle of the night.

After Emily leaves, Dr. Anderson from GF Strong Rehabilitation Centre arrives to assess Simon as a candidate for therapy. The assessment is loud and vigorous, the trim, tiny doctor shouting out instructions: "Squeeze my hand, Simon! Simon, wake up! Wake up, Simon! Push my hand! Make a peace sign!"

And he does. Simon makes a peace sign, with no confusion, and I am very impressed, but after doing this he becomes so tired he closes his eyes and falls asleep, not to be awakened no matter how loud anyone shouts at him.

Dr. Anderson informs us that at this time Simon is not a candidate for rehab therapy and that she, too, is recommending a repeat CT scan. "Simon will remain here," she says, "and we'll continue to follow his progress."

"How long will it take," Lorna asks, "before he is a candidate?"

"It is hard to say." She smiles in the particular way—kind but placating—that I have come to recognize as the precursor to bad news. "For some, it's weeks or months. Some people can take up to a year before they are ready. Some people are never ready, and they are moved to long-term care facilities as opposed to a rehab center."

Lorna and I retreat to the hallway to process this, the news we have been most afraid to hear.

"She says 'long-term care facility' and what I hear is 'life sentence.' I will never let that happen," Lorna says. She is savage in her conviction. "I will never let him go to a place like that."

"No," I say. That night, as I doze off beside Simon, troubled visions of swimming out to the Merry Island Lighthouse thread my dreams: there are deer in the water beside me, and other, more frightening, things. Dark forms, indistinct and obscurely menacing: Bear? Cougar? Sea monster? The lighthouse's flashing white beacon is obscured by a preternaturally thick fog, a fog so viscous and toxic it coats my eyes until I believe I am blind. I keep swimming. I grind my jaw so fiercely that I splinter one of my back teeth, waking with a mouth full of sharp bone.

{ 16 }

TROUBLE IS REAL

AUG 6/08

CT OF THE HEAD

COMPARISON: CT — July 28, 2008, 8:02 pm

FINDINGS:

Left-sided craniectomy is again noted, with brain herniation through the bony defect, which has marginally increased in size when compared to previous films. The right frontal EVD has been removed, without any evidence of acute hemorrhage.

Extensive bifrontal and left temporal contusions continue to evolve, with no new hemorrhage.

Tentorial subdural hemorrhage is largely resolving, and has become hypodense and shifted to occupy the posterior aspect of the left cerebral hemisphere. It only measures approximately 5 mm and exerts only minimal mass effect.

The size of the ventricles is essentially unchanged.

THE TWO HEMISPHERES of the cerebrum are covered in a gray-matter cortex, and each hemisphere is divided into four sections: the frontal lobe, the temporal lobe, the parietal lobe, and the occipital lobe. Within the complex microscopic neural grids of these lobes, thought is translated from an internal symphony of neurobiological electricity to an external world of action and language. I parse my way through Simon's CT scan the way I once did through Milton's *Paradise Lost*. *Brain herniation through the bony defect* means that the cerebral tissue of the left hemisphere has continued to swell outside the hole where Simon's skull was removed, and the brain could be damaged by this contact. *Mass effect* refers to the amount the brain has shifted as a result of an increase in volume caused by the bleeding and swelling. The *tentorium,* or the tent of the cerebellum, is an extension of dura mater that separates the cerebellum from the occipital lobes. That the hemorrhage there is resolving and that there are no new hemorrhages are good signs.

The extensive bifrontal and temporal contusions are the areas doctors continue to refer to as "mushy." I have been given a diagram of the brain with bullet-point lists of each lobe's specific functions. The frontal lobe has a long list, covering a variety of functions and personality traits. Higher functions, we are told—executive functions: the frontal lobes do the work that separates humans from beasts. The list for the temporal lobe is shorter but more critical: memory, hearing, language. I do not like the phrase "continue to evolve," but no one can offer an explanation as to what that means exactly. The important thing, I am told repeatedly, is that there is nothing in this new CT scan to indicate that further surgery is necessary.

The doctor agrees to reduce the morphine to an only-as-needed prescription.

The diagram we were given appears on the following page.

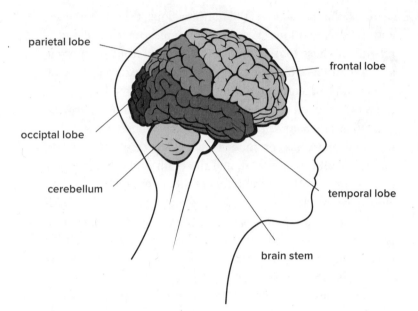

Parietal Lobe

· intellect
· sense of touch
· differentiation of size, shape & color
· spatial perception
· visual perception
· muscle tone, strength & sensation

Occipital Lobe

· vision

Cerebellum

· balance
· coordination

Frontal Lobe

· initiation
· planning/anticipation
· follow-through
· impulsivity
· judgment
· reasoning
· abstract thinking
· smell
· motor planning
· personality
· emotionality
· speaking
· integration of thought and emotion
· self-monitoring

Temporal Lobe

· memory
· hearing
· understanding language

Brain Stem

· breathing
· heart rate
· blood pressure
· movement & sensation for head, neck, eyes, hearing
· relays messages for other movements and sensations

In addition, we are also given a second, more colorful picture in which the various functions of the brain are represented as a cadre of phone-wielding, paper-sorting office workers. Accompanying this picture is a description of how the brain functions, using the analogy of a company. The frontal and temporal lobes, the explanation goes, are the vice presidents. There are many vice presidents, all of whom are constantly receiving information from the rest of the company, and their job is to analyze this feedback, and to use this analysis to inform their judgments and decision making. In addition to this data gathering, the vice presidents of the frontal lobes are responsible for initiating activity, one of the most critical functions performed independently of other brain areas.

The parietal lobe houses the managers, each manager running his or her own department: the mechanics of speech department, a language department, a motor department, a spatial reasoning department, a music department. The managers are in regular communication with one another, as well as with the vice presidents.

At the bottom of both the brain and the corporate food chains are the workers. They are largely unaware of the vice president's big-picture plan. They have straightforward jobs, which they do day in and day out. They remind us when to get a drink of water. They help us stay alert when necessary and, afterward, lull us to sleep so that we can recharge. They express our basic emotions and are responsible for unconscious physical reactions such as the crimson blush when we are embarrassed, the increase in our pulse rate after a fright, the open-mouthed yawn before bedtime, the sneeze after a sniff of allspice, and the tears that flow when we are sad.

In a brain injury, it is as if someone has been fired and the overall efficiency of the company is compromised. *A primary purpose of rehabilitation is to find out who got fired and who is still on the job,* the explanation reads, *so messages can be re-routed and the company can become more and more efficient again.* I read this description over many times, and each time my heart thrills at the words *re-routed* and *efficient;* perhaps healing from a head injury is also an opportunity to pare off all extraneous diversions and reduce the bureaucratic white noise of the brain.

Despite what I perceive to be the optimistic implications of this explanation of brain function, I find the corporate analogy unsettling. The cultural expression of neuroscience reflects the culture the brain exists in; our thinking about thinking provides an interesting window into the time in which we are thinking. If Simon were awake, this would be an interesting point of departure for a conversation we might have: how using the corporate structure of a company to understand our brain so aptly reflects our time.

FRANZ JOSEPH GALL was born in the mid-eighteenth century and became a doctor and an anatomist during a time in Europe when physical features such as facial structure were strongly associated with temperament and moral character. He theorized that development of specific areas of the brain would result in a corresponding development in the cranial cavity and that much could be learned about a person's nature by palpating the bumps and depressions of the overlying skull. People with bulging eyes, Gall contended, possessed strong verbal memory. Those with skull prominence over the ear were excessively destructive and suited for work as, say, executioners. The area on the forehead touched or rubbed by poets during the act of writing represented "ideality." Gall called his theory *phrenology,* and he claimed to be able to identify and localize twenty-seven discrete human faculties, including acquisitiveness, vanity, veneration, wisdom, passion, pride, affection, arrogance, ambition, and a sense of satire, all of which he placed on a detailed map of the human cortex. His ideas were largely rejected by the scientific community of the day but found enormous popularity in the larger culture. *The Constitution of Man,* published in 1827 by George Combe, an intellectual descendent of Gall's, sold over 100,000 copies; at the time only the Bible and Paul Bunyan's *Pilgrim's Progress* were more popular.

The European science community strove to distance itself from Gall's pseudoscience while still embracing the notion of localized function. It was Paul Broca, a highly esteemed French scientist and physician, who made the correlation between a lesion on the left frontal lobe and a patient's inability to speak. Despite widely diverging methodologies, Broca and the phrenologists came to a similar conclusion, locating the ability to articulate language within the frontal lobe. This area of the left hemisphere is still known as Broca's area, the primary speech center of the brain.

Paul Broca's localization legacy carried down through the nineteenth and twentieth centuries. During this time, the field of neuroscience accumulated an immense wealth of data by studying the brain at increasingly greater microscopic levels, breaking down the anatomical, physiological, and biochemical organization of single neurons. It is this legacy that is largely responsible for the first diagram we are given, with its carefully mapped out and segregated regions in the brain.

A DOCTOR ON the ward shows Marc and Lorna the actual CT scan, pointing to the areas where the damage is most severe.

"But," he says, "you can see there is also damage here"—he points—"and here and here." His finger skates over the entirety of the image of Simon's brain.

"Basically everywhere?" Simon's dad asks.

"Basically everywhere, yes. But this area, here, for example"—he points to a smudgy spot on the right frontal lobe—"is midway between two important areas. So, damage there? We don't exactly know what it will mean."

IT IS NOT a comfortable thing, thinking about thinking—akin to attempting to tear the wings off butterflies. In the two hundred plus years since the end of the Enlightenment, neuroscientists have tried to pin down our butterfly thoughts, the how and where and why of them. In direct opposition to the Enlightenment's Doctrine of Equipotentiality, much of the best-known and most influential research of the twentieth century followed in Broca's footsteps. Groundbreaking systems neurophysiologist Miguel Nicolelis refers to this lineage of neuroanatomists as *localizationists*. In the localizationists' framework, the neuron is the single functional unit in the brain; these scientists have worked, and worked brilliantly, to create a map of the brain, placing functions such as language, movement, and creativity in very precise sections of the gray matter's walnutlike folds and fissures. Over the past two centuries, opposition to this idea has been sporadic but passionate. Scientists such as Marie-Jean-Pierre Flourens, Camillo Golgi, the Canadian psychologist Donald Hebb, and, most recently, Miguel Nicolelis himself—a renegade group that Nicolelis refers to as the *distributionists*—have argued that the functioning of the brain is a dynamic interplay of all its billions of microscopic parts and that brain function can only be understood if we look at ensembles or networks of neurons acting in concert, not just at isolated areas of the brain. The dissection of a single ant, the distributionists' paradigm implies, will not provide sufficient insight into the daily complexity of the colony. You can't pluck a single feather, they tell us, and unravel the fluid mathematical abstractions of the shifting, kaleidoscopic flight pattern of a flock of starlings. You can't study a drop of water and claim to understand the sea.

AUGUST 7, DAY 17

Long-term care facility? I can't believe it, Beau. It was such a busy day yesterday, but after everyone left and the evening breeze blew into your room, you seemed to wake up. I told you, as I wetted your dry, chapped lips, that when you were better, some evening soon, we would sit outside at sunrise and eat big chunks of cheddar cheese and drink cold beer and you, in answer, made the rock 'n' roll sign of the beast. You have a new bed that lowers way down so I can pull the cot right beside you and we can sleep next to each other. Every time I drifted asleep last night, you tugged at my pillow until I woke up.

LARGELY BECAUSE OF the legacy of Paul Broca and the ensuing dominance of the localizationists' framework, it has long been believed that language and analytical thought are the domain of the left hemisphere while music and creative expression reside in the right. But more recently, neurobiologists have demonstrated that there is no such strict separation between the processing of language and of music. They have shown not only that there is considerable overlap between the cortical networks responsible for language and music, but that the musical elements (such as rhythm and intonation) in speech play a pivotal role in the acquisition of language: music making implicitly fosters the brain's language network. Given this new research, it is likely that, as both Darwin and Jean-Jacques Rousseau proposed, music was ancient humans' first form of communication, that it is the evolutionary predecessor of speech and remains today a living, breathing, cacophonous bedrock fossil of humanity's varied languages. It is entirely possible that humanity was sharing thoughts, information, and emotion through rhythms and vocalizations long before we even had the word to name it: song.

WHEN SIMON WAKES, he indicates with his fierce right hand that the tape and nose tube are irritating him. I tell him I will speak to a doctor or nurse just as Karen, the occupational therapist, arrives with Marc and Lorna. Karen asks Simon if he understands what the nose tube is for. Although it has been previously explained to him that it is there to provide nutrition, he signals with a waggle of his hand that he is unsure. Karen re-explains, and he squeezes, yes, he understands. She has brought a blue duotang labeled *This Log Book Belongs to Simon,* where we can all mark down progress, observations, or concerns. As I leave to go to the hotel for food, a shower, and a nap, Simon is being transferred into the bed-chair.

While Simon is positioned in the upright chair, eyes closed and head slumped forward, Marc continues singing. He sings Gram Parsons's "Hickory Wind," an urgency threading through the gentle melody.

At the start of the third verse Simon lifts his head and looks toward his father. As Marc sings "It's a hard way to find out that trouble is real," Simon, for the first time since his fall, mouths the shape of words. Although he makes no sound, he is, in effect, singing the lyrics along with his father.

Later, when I return to his room, Simon gestures "you" (pointing at me) and "talk" (making a gabbing motion with his hand), then points to his nose tube.

"Yes," I say. "I spoke with the nurses. Do you remember what Karen explained about the nose tube?"

Yes, he nods and mouths "thank you" and "love you." I think. I think that is what he is saying. I want so much for that to be what he is saying.

"I love you too, Beau. You know I do."

He pushes the right side pillow off the bed and gestures emphatically for me to sit beside him. I sidle into the narrow space between his body and the lowered guardrail, massaging his left hand and wrist, as he falls back to sleep.

SANTIAGO RAMÓN Y CAJAL, a fiercely free-thinking Spaniard, was awarded the Nobel Prize in 1906 for his work in neuroanatomy. Ramón's work demonstrated how axons grew, and he argued forcefully that nerve cells are singular and independent elements. The idea that music training can produce anatomical changes in the brain can be traced back to Ramón when he argued, "Everybody knows that the ability of a pianist . . . requires many years of mental and muscular gymnastics. To understand this important phenomenon, it is necessary to accept that, in addition to the reinforcement of pre-established organic pathways, new pathways are created by the ramification and progressive growth of terminal dendritic and axonal processes."

This is what music demands: that new pathways be created.

AUGUST 8, DAY 18

Big day, babe. At 6:30 this morning I went to write an update in the log book. You gestured at the book and I handed you the pen and held up the back page and you wrote "Logbook." When I asked you to write your name, you printed out "Simon." Later, while I was discussing the need for an ear specialist to take a look at your ears and the nurse was disagreeing, you gestured for the log book and wrote "take Kara's opinion." "Opinion" took several tries and you didn't spell it quite right, but close enough.

Seriously!

I have so much faith in your ability to heal, but, honestly, I didn't think we would reach this point so fast. I was so excited to show your parents, but I could see the look of dismay on your mother's face when she saw your spidery, barely legible scrawl. This is it? her look said. This is the big breakthrough? She is terrified at how slow your process of waking up is. She'd prefer (as would we all) that you'd just sit up and start beating her at Scrabble. Right. Now.

Still... it has been such an exciting day with you writing. It's hard for me not to get too ahead of myself. I, of course, wanted you to keep writing all day. This morning you asked, "What is reason for stay?" and I explained that you had fallen and hurt your back and head. You also wrote "general information for the persona" and, when Guido arrived, "A blank page for Eli...," two messages which made me feel especially pleased. "Simon has some places to rock!!" was perhaps my favorite message of all.

You did continue to write messages with vehemence—underlining and punctuating words with several, determined period marks— but they lacked the clarity of your earlier notes. The lines you drew looked like letters, almost. I felt as if you knew exactly what you wanted to say but the message was getting lost in translation.

But every day is a roller coaster. I left at 4:00 to eat something. Your mom said shortly after I left, your fever spiked and your hand movements became agitated. You were chilled and shivering, the fever was so intense. There has been a lot of dense bloody phlegm being suctioned out of your lungs. The doctors are also worried about a possible urinary tract infection. A nurse just came and took bottles of your blood (nice bottles, it looked a fine vintage). We'll figure out what this is.

It's 9:00 now. I sense we are in for a long, difficult night, but that's okay. Eli is coming in tomorrow. Everything will be better with him here.

AUGUST 9, DAY 19

AS THE SUN rises, I slump into the chair beside Simon's bed, my sense of urgency in recording these pivotal days supplying me with just enough energy to update my journal. I have been up all night, along with the overnight nurse, filling small white bags with ice from the ice machine to pack around Simon's body, tucking them into his armpits, over his forehead, and around his groin. His fever has hovered around 40 degrees Celsius all night, and because his liver is showing high levels of toxicity, he is unable to take a Tylenol. Ice bags are the last line of defense against this raging fever. He is so hot, it takes less than ten minutes for the ice to melt enough that it needs to be replaced.

Simon wags his hand at me, signaling he wants to write something. I pass him the pen and hold up the log book so that he can scribble his message, but the words he writes are undecipherable. The only thing I can clearly make out is the phrase "I Simon" and the word "wait."

"I'm waiting, Beau," I say. "You take your time. As much time as you need. Do you remember why you are here, in the hospital?"

Simon's right hand, fingers and palm flat, tilts from side to side, a gesture we have come to interpret as indicating uncertainty. I give him my standard explanation that: one, he hurt his head and back but is in stable condition; two, he needs a lot of rest while the swelling comes down; and three, he can trust his body because he is amazing and is getting better every day. I have been repeating this explanation to him, using the same words over and over in the hopes that he will begin to remember what is going on from one waking moment to the next, and this time I am rewarded by an enthusiastic thumbs-up.

Although his forehead is warm and clammy against my palm, it is clear the fever is lessening. I play Gram Parsons's "One Hundred Years from Now" on the small boom box Marc has bought. The look on Simon's face suggests that while the song is familiar, he can't quite place it. He mouths (I think) "What is this?," suddenly exhausted. I switch it off just over halfway through, replacing it with Glenn Gould's recording of Bach's Prelude in C Minor. This music is more soothing, and Simon drifts off, finally, into a peaceful sleep. When Marc and Lorna arrive, I retreat to my darkened hotel room.

9:35 P.M.

Beau! Eli was here today and described how Sully had dared him and his friend Nate to wear girl's short shorts to play tennis in. Eli said he and Nate had accepted the dare but were worried about "their junk hanging out."

And you laughed.

Not just a smile but a full—albeit noiseless—movement from the belly.

Eli wrote you a note that I am to show you every day he is not here: Hey Man, Stay strong. Everyone is cheering for you. I love you. Eli.

Your fever is heading up again.

The nurse—another Judy—gave you a bed wash and then she said, as she pulled on a pair of gloves, she was going to get some lubricant and "perform your bowel care." What a look on your face. Oh babe. I am glad that you are likely forgetting all of this. The look was one of questioning horror. All I could say was something dumb about this being the best way right now and telling you again that this is transitory—moments in time—that we will be moving through and past this.

I try to imagine what it must be like waking up and not remembering the previous day. I do feel like there is a bit of a cumulative buildup of information for you. Your confusion today is more articulate, less abstracted.

AUGUST 10, DAY 20

IT IS ANOTHER sleepless night replacing ice bags over and around Simon's scalding body, but by 6:30 in the morning the fever seems to have finally broken. I place rolled-up towels under Simon's hands to facilitate straightening the fingers. The left hand, still unmoving, remains tightly clenched.

As Simon wakes I play the Glenn Gould CD until he signals—with a definitive slicing motion—*enough*. I ask him if he would like me to read and I pull out the Cormac McCarthy book. I have only read a few sentences when I am interrupted by a whispery squawk. Simon is trying to speak.

"Sorry, Beau," I say. "I'm not sure what you are trying to say."

He vigorously waves his hand at the log book, and when I hand him the pen he writes, "Why am I alone riding up Coast."

"I'm not sure I understand the question," I say. "But I can understand that at times right now you must feel very alone. But I'm here. So are Eli and your mom and dad. And so many other people: Emily, Sarah, Rob, my mom, Guido, Joe, Sully. If not always in person, then in spirit." I explain to him that after he fell, he rode in the Air Evac, alone, to Vancouver and that Guido met him at the hospital and that I came as fast as I possibly could.

Simon's right hand drops the pen to tug at his blue hospital dressing gown. He shocks me by vocalizing clearly: "What will I wear?" I can't help but laugh. It is a most ironic first statement, given Simon's usual disdain for all things sartorial. Because of the trach, his voice is thin and mechanical-sounding but ... it is his voice. Talking!

"I need clothes." This next statement is more of a whisper, and it is obvious that the effort to push enough air to vocalize is exhausting.

When the respiratory therapist arrives, she covers Simon's trach opening. His breath is strong and he graces her with two words: "Hello" and "Perfect." After the nurse suctions out his lungs through the trach hole, he says, "That was remarkably uncomfortable," carefully enunciating each syllable in a manner of speaking that reminds me of his grandfather.

He laughs several times at my dumb jokes—big, full chuckles—and, when he sees my cup of tea, asks what I am drinking. He remembers the explanation from yesterday that he couldn't have food or water by mouth yet and that he was getting nutrition from the damned nose tube. He smiles when I tell him I will cook anything he wants to eat when the tube is out. Later, when I ask if he needs anything, he says "Coca Cola," and I say the best I can do right now is honey lip balm, and once again he laughs.

9:30 P.M.
Well, Beau . . . when you woke up this morning, you really woke up. You have been awake all day, connecting with the nurses, being grilled by your mom (she ran you through your paces—identifying colors, numbers, letters, geometrical shapes and reading a clock face—all of which you answered like a pro) and engaging with visitors.

Al and Julie came to visit, and you, without being prompted, remembered that you had been set to play a wedding gig in Vancouver yesterday with Al (and Pat and Kristi). You asked Al how the gig went and who played guitar and apologized for not making it. You remembered the Olympics were on and told me you wanted to watch some "goddamned sports." You agitated for getting a TV hooked up early in the morning and remembered to follow up throughout the day, badgering me every time I walked into the room, so insistent that I have to override my anti-TV sentiments. Nurse Laurel obliged us and brought in an ancient idiot box in the evening. No cable yet, but we watched half of an old VHS cassette— Girlfight—before you fell into a deep sleep.

It has been one of the best days of my life.

AUGUST 10, DAY 20

THIS IS THE first day Simon will remember since his accident. Although many questions about the brain injury still remain, some of the largest are broadly answered: significant aspects of his memory, language, and personality are still amazingly, gloriously intact.

{17}

A MYSTERY GIFT

S SIMON WAKES, like a bear from hibernation, so does his appetite. He is famished, and all he wants is solid food. Well, he wants two things: food and access to round-the-clock coverage of the Olympics. This hunger for both nourishment and entertainment means that not only is he alive, he is Simon. The weight of all those dangerous, treacherous thoughts I worked so hard not to even think—What if Simon didn't wake up Simon? Who would he be? Would he be angry? Would he be desolate? Would he be lost or trapped within the bruised and caved-in parts of his brain? Would he still love me? Would he still want what he had always wanted before?—is momentarily suspended. The profound normalcy of his requests verge for me on the miraculous. TV and chicken shawarma.

The TV proves easier to provide than food. The day after Simon's awakening, he is moved out of Step-Down and into the spinal cord ward proper. Marc fills out a request at the nurses' station, and, shortly after, the cable guy hooks up a small TV in

Simon's room and he is able to watch swimmer Michael Phelps take the men's 400-meter medley, the first of Phelps's historic eight Olympic gold medals.

Although I have explained the nature of the spinal cord injury to Simon when he learns he will be transferring out of Step-Down, he writes this note: *If I'm moving to another room I'll need a set of crutches to assist me to!!*

None of us know how to respond. Heartsick, I realize we are all going to have to experience the loss of his legs many times over before it truly sinks in.

"A stretcher is best for today," I say.

"Okay," Simon agrees, without complaint.

"Maybe we'll check out crutches at a later date," I say.

"Okay," he says. Again with no complaint or resistance. It is unlike Simon to be quietly compliant. I wonder if it is the head injury that makes him so mild and even-tempered. Or is it some combination of shock and self-preservation? An internal safeguarding mechanism? One that insists on tranquility and moderation and that allows him to grasp the extent of his injuries in manageable increments.

MARC UPDATES THE log book with the news of Simon's transfer out of Step-Down, writing that during the move "Si was full of humor and vitality." I love this entry because that is what I see most in Simon too—his humor and vitality—though I don't think it is readily apparent to the world at large.

Often Simon can stay awake for no longer than ten or fifteen minutes at a time. The only part of his body with voluntary movement is his right hand. He is now breathing on his own, but the trach is left open in case that suddenly changes. The suctioning of his lungs stimulates a cough, but it is weak and he cannot swallow. His head lies on the pillow inclined to the right, and only with enormous effort can he roll it toward his body's midline. He is unable to move his eyes to the left, past that midline point. At its best, his voice is clear but flat and robotic, and it takes a titanic effort to move enough air to create an audible sound with the trach still in. Most often, Si's previously booming voice whistles and croaks out a thin, gravelly noise. The left side of his face droops, and there are large and mushy pressure wounds on the back of his head and his right foot. Because he is still fighting recurring bouts of pneumonia and various internal infections, his fever rises in the late afternoon and then he becomes even more sleepy and disoriented. Still, there isn't a more apt phrase than "full of humor and vitality" to describe his disposition.

SIMON'S NEW ROOM on the spine ward is large and single occupancy, and for the first time he is allowed plants and flower arrangements. Lorna buys a potted jasmine to sit on the windowsill, its sweet fragrance an antidote to the antiseptic smells of the hospital. Beyond its white petals and waxy green leaves, the view from the window overlooks the emergency helicopter landing pad.

"Isn't it cool to see the helicopters?" a student nurse asks Simon. He is young and pleasant, eager for feedback from both Simon and me, but the tone of his voice—verging on cutesy and condescending—grates. A red-hot filament of fury sparks and I grip the guardrail on Simon's bed, counting to three before I speak.

"Maybe not so cool when you're rushed here in one," I say, pleased to see the look of apologetic horror cross his face.

"Whatever," Simon mouths, shrugging his right shoulder.

I am the one, not Simon, traumatized by the comings and goings of the helicopter, each arrival or departure intimating a potential catastrophic injury. Each time the helicopter lands, I think some devastated family from Kamloops or Kelowna or

Prince George is about to embark on a journey similar to ours. It is overwhelming, all this potential hurt, and I wonder how the doctors, nurses, and rehab staff face it every day. Simon, however, finds the helicopters mildly interesting; more disturbing to him is the high-pitched beeping of the IV fluid line whenever the tubing kinks, which is often. The IV inserts into Simon's right hand, which, as the only mobile part of his body, is often in ceaseless motion, even while he sleeps—as if that one limb serves as a repository for all the vast amounts of energy of his pre-accident body. The insistent squawk of the alarm is loud enough to drag him, disoriented and dozy, from a deep sleep during the night. When he wakes, I explain to him where he is as I unkink and restart the fluid line. I repeatedly suggest that his right arm should try to relax along with the rest of his body.

"Impossible," he replies one night. "It is busy playing a John Hiatt song. That arm has its own presence."

During the long nights when Simon is disoriented or drifting off to sleep, I repeat the message I have been delivering since his time in the ICU. That his injuries are serious but that he is stable. That he is safe. That he is strong. That we will be patient and let the swelling subside. That he should trust his body and love his body because his body is amazing. That his body knows exactly what to do. I believe what I say, but even more I believe in believing this: I want Simon to feel positive and empowered on a bone-deep, cellular level. This approach, I believe, will give him the best possible chance for recovery.

"Thank you," he says one night. "I'm going to set up Redroofs Road for you, Kara, so it will go with the flow of your bones."

And another night: "I know, babe, I know. I need a mystery gift for you, or whoever is dealing with the potatoes."

And each night I scribble down all his beautiful nonsense,

each message a lyrical and musical map that gives me insight into the way that Simon is rebuilding his brain.

He is actively rebuilding, a process he clearly articulates when a speech therapist questions him about his ability to think, to find and to speak a specific word. Did he find it difficult? Was the word he thought the one he eventually spoke, or was it a different one?

"It's like I have an idea," Simon replies in his slow, robotrach voice, "but I can't just spit it out. I can't just go A to B. I need to take the long walk around the park to find all the words I need."

"Well," the speech therapist says, "if you are doing that on your own, then you don't need me."

MARC AND LORNA and I maintain our established routine: I spend the nights at the hospital; they spend the days while I nap in the hotel room and start to reorient myself to the real world of paying bills, balancing budgets, and planning for our still very uncertain future. Grade 11 starts for Eli in a few short weeks. Simon is getting stronger, meaning that, sometime within the next year, we will return to the coast. Our Hobbit house on Redroofs Road won't accommodate a wheelchair, so we will need a new place to live.

Simon begins the hard work of learning how to swallow and to reawaken his left hand. We start to take inventory of his injuries, addressing some of the previously overlooked non-life-threatening issues. His ear canals need to be cleaned of the dried blood that has accumulated, and his hearing, especially in the right ear, needs to be tested, but for this to happen he first needs to build up his tolerance for sitting in an electric wheelchair from the ten minutes he can now manage to at least the hour it will take to go to the Ear, Nose and Throat Clinic at the Diamond Centre, adjacent to the hospital. He has no sense of smell. This could be due to nerve damage, or perhaps damage to the olfactory receptors in the cerebral cortex. The doctors refer to Simon's left-side neglect as a stroke, explaining that it is most likely caused by something called Kernohan's notch phenomenon. At the time of the accident, the bleeding in the left side of Simon's temporal lobe shifted his brain so far down and to the right that a portion of the right side of the brain stem called the cerebral peduncle was pressed against the hard structures of the skull, creating a notch, an area of damaged tissue. The right cerebral peduncle contains motor fibers that cross over to the left side, and damage to these motor fibers results in hemiplegia, or left-side paralysis. These fibers are so densely packed in the brain stem that even a slight amount of damage can cause catastrophic results. Brain stem injuries are more difficult to recover from, though it is not impossible for neurons to reroute. Over time, six months to a year, Simon might recover some hand movement. Or he might not.

"It's hard to say exactly," the doctor summarizes, a phrase we hear often.

Each day, in the morning and late afternoon, Marc, Lorna, Simon, and I regroup for a debriefing on how the day has gone. After so many days of waiting, waiting, waiting, it is a relief to finally have something to do.

Such relief during these days! And still so much fear. Such joy and such sorrow. Never in my life have such extremes of emotions so naturally coexisted. Like unruly children, they trip over each other's heels, finish one another's sentences. Simon is alive, yes, but malnourished, dehydrated, paralyzed, too weak to move. Alive, yes, but with a mushy, weeping wound on the back of his head, a skull cracked open like Humpty Dumpty, and no ability to control his bowels or bladder. Alive, yes, but what kind of life lies ahead?

The thought of our future life is impossibly daunting. The future, like hope only a short while ago, is too precious a commodity. Right now, moment by moment, I am capable only of facing the most imminent challenges of the day. It is too big, too overwhelming, to think about the future.

ONE DAY WHILE my brother is visiting, Simon asks him how long we are staying in Powell River (the town where my mother lived) while simultaneously acknowledging that he is looking at the Vancouver skyline. Simon is awake, seemingly lucid, and yet disoriented. He acknowledges the inconsistency in his logic when it is pointed out, saying, "It's weird. I had it in my mind we were in Powell River." This episode causes a jolt of panic for Lorna, but I am not too concerned; I believe that somehow his present confusion relates to the confusion he experienced at the time of his accident. He doesn't remember the accident or the helicopter, but one of his earliest written questions was "Why am I alone riding up Coast." He had been losing consciousness as Emergency workers loaded him into the Air Evac, and I believe that somehow he thought he was being taken to Powell River, north of Halfmoon Bay. I had spent much of the previous year in Powell River, at the hospital with my mother for her chemotherapy treatments. Simon, even as he was losing consciousness, was planning to meet me there.

Most of the time Simon is rooted in the world we were living in before the accident—a world that now feels lost to me. Did the Precious Littles play the Islands Folk Festival? Did someone fill in on guitar? How did the gig go? How are we going to get to Chilliwack in two weeks for Eli's annual soccer tournament? Where will we stay? Could I call Dave on his behalf? He arranged to buy a piano from Dave for my birthday—sorry to spoil the surprise!—and I need to call Dave and tell him we still want it but won't be able to pick it up right away. When Ray, the owner of StraitSound recording studio, comes to visit, Simon emerges from his dozy state and immediately apologizes for not showing up for the recording work they had booked for that week.

"I'm not going to be able to make it, man," Simon croaks through his trach hole. "Do you have someone else you can call?"

"That's okay, Si," Ray says. He brushes errant tears from the corner of his eyes. "I'm taking a little time off work right now."

Simon smiles. "Me too," he says.

WITH THE HELP of Karen, Simon's pragmatic and personable occupational therapist, we set up Simon's room to help him orient toward his left side, moving the visitor's chair, my sleeping cot, and the boom box to his left. I encourage Simon to hold a washcloth in his left hand, and with my hand over his hand, help him make small cleaning swipes over his chest and face. When he is relaxing, we all spend hours massaging that left hand, manipulating the wrist and fingers, and I place rolled-up washcloths underneath his palm while he sleeps to encourage his clenched and curled fingers to release and straighten. Small, beautiful movements gradually begin to return to his hand, arm, and shoulder.

"SUCH A GENTLE and yet such a strong man," Nurse Zola says to me one day as we walk down the hall toward Simon's room together.

"Yes," I say, and start crying, as I often do. Outside Simon's room, tears can come quickly and unannounced, shocking in their intensity and force. I have never cried easily, but now it seems I cannot stop. I am snotty and slosh full of tears, tears that are a confused expression of the currents of sad-happy, relief-regret, hope-despair that constantly churn my insides. Most often I don't even know why I am crying, but I do know now. Zola has described the essence of Simon. What, of all his many excellent attributes, I love the most: the ease with which he embodies seemingly contradictory traits. He is gentle and strong. Fierce and playful. That this, even in his compromised state, is what shines through to a relative stranger fills me with a kind of devastating happiness, an emotion too immense to put into words. I can only fall into it and allow it to momentarily swallow me. I huddle against the wall outside Simon's room and bawl like a baby, choking and huffing for air while Zola gives him a shave, finally ridding him of his hideously overgrown beard. When the storm passes, I go back to the room. Simon smiles from the bed with newly shaved head and face—a wide-open grin. Without the beard, the amount of weight he has lost is more evident, as is the large area of his head where, underneath the skin, the skull bone is absent. Still, without the beard he easily looks ten years younger.

"You're rocking the sexy monk look," I say.

Zola winks and says, "I'll leave you two alone."

Zola has worked with spinal cord patients in many countries and is a font of information. She and Simon establish an immediate rapport. Simon actively works at cultivating good relationships with all the nursing staff. Zola is a favorite, as is Corrine, a Jamaican national, who, the day when Usain Bolt obliterates the men's 100-meter track record, spends a little extra time in Simon's room celebrating.

"He was barely trying," Simon crows from his bed.

"That man is living proof," Corrine sings, "that Jesus loves Jamaicans!"

Often, though, when new nurses arrive at the beginning of their shift, they address me, asking questions about how *he* is doing. I rarely have to redirect them to ask Simon himself, however, as Simon, if he is awake, will interject:

"I'm Simon," he says, in his most assertive yet gracious voice, "And you are...?"

The way Simon handles these interactions impresses the hell out of me. He has to constantly assert his selfhood to medical workers who often defer their queries to me or his parents. As a previously physically strong, very physically present person, he now has to find that sense of selfhood without the use of his body. Since he is recovering from a head injury, he has to do this while in the process of reassembling himself. As someone who is suddenly utterly dependent on a stranger's assistance for the most basic of bodily functions, he has to engage in these relationships from an oddly imbalanced interpersonal dynamic. And he does. Not only does he do it, he does it without anger or resentment, without complaint or shame or defensiveness. His "Si-ness," which I had felt even from the depths of his coma, is immediately present when he awakes, independent of the physical functioning of his body. This, to me, is impressive—inspiring.

The word *inspiring* tends to be used in a vague, general way when the able-bodied comment on somebody with a disability, so I believe it is important to be precise: What impresses me is how effortlessly Simon's sense of selfhood remains intact, independent of so many things we typically take for granted, and how effortlessly he asserts it when confronted by a medical system whose first introduction to him is with the label "traumatic brain injury patient." Inspiration is located in a very precise moment—"I'm Simon, and you are...?"—and it convinces me, compels me, to work at being a better, stronger, more internally intact human being.

RETURNING TO SOLID food is an agonizing process for Simon. He must undergo multiple "swallow" tests in which he takes small sips of food of various consistencies: pudding, a fruit puree, and thickened water, each colored with a blue dye. A few hours after the test, when Simon's lungs are suctioned through the trach hole, blue dye colors the phlegm, indicating an inability to co-ordinate the complex acts of breathing and swallowing: every time he swallows, particles of the blue food and liquid are aspirated into his lungs. With this type of uncoordinated swallowing, choking and the development of pneumonia become potential hazards. Simon is not given clearance for solid food. He is starving and dehydrated, and when his hunger becomes overwhelming he drifts toward despondency. Eventually he is given a swallow test while being X-rayed and receives clearance for everything but liquids. It is a triumphant moment when he is able, finally, to dig into a tray of blended beef and peas, pureed pears, and a glass of "water"—a gelatinous lemon-infused monstrosity thickened with cornstarch.

As the days pass, there are small increases in movement in Simon's left hand and fingers. His head can now easily roll toward the midline of his body. With difficulty, he practices tracking his eyes, but every time his line of vision passes to the left of midline, his vision doubles, even quadruples. Damage to the optic nerve, we are told—a permanent state. But days later, this too improves to the point of being a non-issue. A physiotherapy schedule is established. The physiotherapist, Brian, slides a mesh sling under Simon's body and attaches it to a hook suspended from the ceiling; when a flip is switched, the sling hoists Simon into the air and Brian guides his body to the wheelchair the hospital has provided. Initially, the entire physio workout consists of Simon's surviving this difficult transfer, holding his big head upright, and not falling asleep. Simon, despite the staggering challenges of navigating this situation, remains playful and positive. He gives Brian the nickname "Chuck" because of his passing resemblance to UFC fighter Chuck Liddell. After a few days, Simon has progressed to attempting some exercises in the physiotherapy room. His left arm is strapped onto a skateboard-like contraption, which he is asked to move in a figure-eight motion across a tabletop.

He stares at the arm, sweat breaking out on his forehead, willing it to move.

Nothing happens.

He stares some more. Then, grinning, he reaches for his left forearm with his right hand and pushes the skateboard, which rolls an inch or two.

"Cheater!" Brian laughs.

"I'm trying, Chuck, I'm trying," Simon says. "Don't give up on me, man."

SIMON IS CONSTANTLY being assessed. He is—according to various therapists—too swift and impulsive when eating, gobbling a whole plate of food down in minutes. Cognitive tests indicate he has deficits in his executive functions, the more abstract and complex functions of the frontal lobes such as problem solving and short- and long-term planning. He is, perhaps, too cheerful and positive given his current situation. When an occupational therapist explains to Simon that the hospital will lend him an electric wheelchair until he has one of his own, he politely says no thanks.

"I prefer crutches over a wheelchair," he adds—evidence of his general lack of awareness and acceptance of the severity of his injuries. But I see it also as a statement of his desire for independence. Of course he prefers crutches over a wheelchair. Who wouldn't?

The doctors, nurses, and rehab health care workers we meet at the hospital are almost always professional, compassionate, and generous with their knowledge and skills, and they provide us with a wealth of information. Still, I often resent this atmosphere of constant assessment. Rationally, I know these assessments are necessary and are in Simon's best interests; emotionally, they make me feel both enraged and defensive. I know, after years of working as an advocate for adults with various disabilities, how quickly a label can be applied and how limiting and self-fulfilling one can be. I know that these skilled and most often wonderful people are neither wrong nor ill intentioned. They are doing their job and they are doing it well, providing us with their reports, their recommendations, and handouts from various information resources. But it is my job to integrate that information into a larger perspective. Many of the assessments could, with a slight shift in focus, be reframed to reflect what Marc, Lorna, and I see in Simon: his humor and vitality. He has always wolfed down his meals. He isn't necessarily impulsive; he is, as he says, fucking hungry.

I am not in denial about the severity of the injuries. His injuries are undeniable. But it seems unwise to rush headlong into naming and cataloging Simon's "deficits" and preparing for a lifetime littered with various "compensations," two words we are becoming all too familiar with. Unwise, even, to rush into facing everything all at once. The brilliant surgeons, ICU, and Step-Down staff have given this to Simon: a lifetime of learning how to accept and deal with his injuries. Dr. Griesdale had explained that three months was a likely timeframe for Simon to be in a coma. But although it feels like it's been a lifetime, the accident happened less than one month ago. Simon doesn't need to rush into anything; he is already way ahead of schedule.

I make a conscious decision not to worry about or obsessively evaluate his executive functions. Not yet, anyway. There are plenty of other, more pressing things to worry about. I worry about his elevated liver enzymes, spikes of fever, and dehydrated, collapsed veins. I worry about his description of his right ear: he says he can hear blood flowing through it but nothing else. And I worry about his left hand.

They are small miracles, human hands—no tool is more perfectly crafted for music, art, or love. A working left hand means, for Simon, the ability to fret the strings on his guitar.

"Love the left" becomes my catchphrase as I work my fingers into his cold, swollen palm and clenched grip, channeling all my impotent worry, pouring whatever strength remains in my jittery and nervous soul into the small miracle of his left hand.

"WHO," SIMON DEMANDS as I settle into the cot beside him, "gave Lorna a book on brain injury?" It is early evening and his parents have left for the day.

"I gave her the Jill Bolte Taylor book," I say. "And Sully and Patricia brought some type of head injury guidebook. British publication, I think." Lorna has been running what she terms "classes" during the day, in an effort to push Simon to open up as many mental files as possible. It seems to me that most of these files—language, memory, spatial recognition—are opening on their own, but Lorna's classes provide a lot of important stimulation for Simon, and she is the only one who has the energy to approach these activities in such a focused and directed way. Beside me, Simon groans with an aggrieved air of exhaustion.

"Why do you ask?" I say. "You okay?"

"I'm tired," he says. "Lorna should never get ahold of medical books."

"One book and she's an expert, right?" Now that Simon is awake, we are all a little guilty of this, hastily cobbling together a Wikipedia medical degree, a modern phenomenon that must drive real doctors crazy.

"Well, then," Simon snorts, "I think somebody must have given her a book on being a kooky-assed witch."

I can't help but laugh. It is Simon's first expression of anger, and it contains his familiar bite. It is fitting that it is leveled at his mother, mothers being the safest place for children to lodge their disgruntled complaints about the inadequacies and stresses of the world. The person I would most like to share this flash of snarkiness with is his mother, because, of all of us, she is most worried that his manner is too easy-going, too accepting and complacent. It would be a relief to her to see a little anger or frustration on his part. But I worry at how she will receive his complaint. We are all fragile right now, and it might hurt her feelings. She and Marc have willingly joined Simon and me in this small, surreal fishbowl of medical trauma, and I know being here is of the upmost importance to her. I wonder if she sees, as I do, how much their continued presence means to Simon. How he absorbs every moment of his parents' love and undivided attention, how it causes him to swell and settle and radiate a joy that is not childish but childlike in its purity. I decide it's best for me not to interfere and say anything about his cranky outburst. Plus, I don't want her classes to ease up or stop.

AUGUST 16

ON A SUNNY Saturday afternoon, a "Bring Simon Home" benefit is held at the Roberts Creek Community Hall on the Sunshine Coast. The event is spearheaded by our friends Sue Lasby and Patricia Van Der Veen, though many, many people donate time, money, and services to make it happen. Local bands and musicians put together a full day's lineup of music, and a silent auction is organized. Marc and Eli attend the event, and I send an open letter to our community thanking them for all their support—emotional, logistical, financial—which Guido reads out loud.

Marc breaks down and cries, overwhelmed by this public show of love for Simon. It is a difficult day for Eli too, difficult to see his normally steadfast grandfather cry. Difficult to talk to so many community members about what is happening at the hospital when what he wants, more than anything, is for it not to be happening at all. Difficult and yet also amazing, because as much as he wants to run away and hide, he is embraced and loved and held in place.

"That kind of love," Lorna says when Marc returns to the hospital and gives us a full account of the day, "it shows what kind of person Simon is."

"Yes," Marc agrees, then adds, "But even more, it shows what kind of community the Sunshine Coast is."

It is true: the Sunshine Coast is a remarkable place. The support of our community works through me in a profound and transformative way. The amount of money the community raises is significant and ultimately will help in so many critical ways over the coming year. But, even more importantly, what our community gives us at this time is hope: real hope for a future with choices. This is no small thing. At a time when my most negative internal monologue is focused on how bereft I am of the resources, both personal and material, to deal with loss and change of this magnitude, our community makes me feel as if we are the richest family in the world.

All Your Junk And
Your Catalog Too

MY EARLY MEMORIES of waking are patchy and jumbled. I was sleeping so much and so deeply that each time I woke, it felt like a new day, so events I recall as having taken place over the course of a week actually happened in a single twenty-four-hour period. Even though I am told that I had been asking questions about my legs and that the nature of the injury had been explained to me, the first moment I recall realizing that my legs were paralyzed was when two nurses were transferring me from my bed onto a stretcher. They manually rolled my body onto its side and placed a long wooden board under me and then slid me over to the stretcher. I saw the nurse at the bottom of the bed grab my legs and lift them but I felt nothing. They were legs that could have belonged to someone else. That's the first time I remember consciously realizing that I was paralyzed.

I also vividly recall the first shower I had post-accident, at the end of my stay at Vancouver General. A hulking male attendant wheeled me into a roll-in shower, soaped me, and rinsed me down. The warm water felt great. Afterward, I thanked him for "washing all my junk."

"No problem, man," he said. "I washed all your junk and your catalog too." To this day I still don't know what a "catalog" refers to!

When you are in the hospital, what Kara referred to as "the indignities of the flesh" are the norm. People see you nude and vulnerable. They wash you and perform "bowel care," which, no matter how kind and skilled the nurse, is exactly as invasive and scary as it sounds. You have to get used to strangers taking care of your body in a very intimate way. God bless 'em. It's a very heavy job, that one.

{ 19 }

MOMENTS IN TIME

WHEN SIMON'S LUNGS no longer need constant suctioning, his cough reflex is strong, and his breathing is deep and regular, a duo of doctors arrives in his room to perform a decannulation, removing the trach tube from Simon's windpipe. The hole in his throat is covered with a simple gauze bandage; it is a wound, one of the doctors explains, that, for the majority of people, closes up and heals with almost magical speed. After this, the move to GF Strong happens very quickly. A bed is available there, so the rehab doctor returns to VGH to reassess Simon. There are concerns that Simon's health is not yet strong enough for the rigors of rehab. He cannot tolerate being in an upright position for more than half an hour, and his IV antibiotics have only just ended. But this interview is much different from the first one. The rehab doctor asks Simon only one question:

"Do you think you're ready for rehab?"

"Yes. I want to go," Simon says, measuring each word carefully so that there is no confusion in his meaning. "Staying here means staying in the same place. I need to keep moving forward."

THE NIGHT BEFORE the transfer to GF Strong is a sleepless one. Simon has butterflies—the kind, he says, you get the night before starting a new school. On the morning of August 21, before the transfer, Simon and I are accompanied by an attendant over to the Diamond Centre to get his hearing checked. The results of the test are devastating. The hearing loss in Simon's right ear is 100 percent, and it is irreversible. Hearing aids or cochlear implants would be of no help. His right ear is gone. This, I think, will be the diagnosis that breaks Simon. I feel it breaking me.

"It's tough to get that news," I say as the attendant wheels Simon away, through the underground parking lot, past the sign for the morgue and the tissue bank. Despite the heat of the day, I shiver at the thought that the rest of Simon's skull is in a fridge somewhere down here. I reach for his hand. "How you doing, Beau?"

"Well," he says, "it sucks, but... one ear won't be too bad for studio work. I can master in mono now."

I don't exactly understand the benefits of mono mastering, but it doesn't matter. He is still smiling.

"WHAT CHANGED?" A doctor in the Infectious Disease Unit at St. Paul's Hospital poses this question to Simon and me two years after he is discharged from GF Strong. She is compiling a case history for Simon, and when we finish describing those eleven days between the two interviews, she is curious. "What changed between the first rehab consult and the second one?"

Three things pop into my mind. "We requested the morphine drip be stopped," I say. "His fever eased up. And we started playing Bach in his room. Glenn Gould playing Bach.

But I don't think those changes made the critical difference. I think at that point it was just a matter of time."

So . . . what *did* make the critical difference?

So many moments contributed to making a difference, starting at the very instant of Simon's accident. His co-workers acted efficiently, calling for paramedics and insisting even before the ambulance arrived that an Air Evac was required, ensuring that, despite our distance from the hospital, Simon was in an operating room just over two hours after his fall. It made a difference that we live in a country where health care is universal and that the quality of care is, at any given time, outstanding, especially so in an emergency department or intensive care unit. It made a difference that Dr. Haw and his team weren't off that day or having lunch or in the middle of another surgery when Simon arrived.

I wrote in my letter to the community that its support made all the difference too: I believe that now as I believed it then, though there is a qualitative difference in my belief. Now, the idea of community "love" and "support" has drifted back into the more abstract realms it generally occupies. It was different during the time Simon was in the ICU: I experienced that support as a vital, physical force buoying me up. It is a word—*buoy*—that surfaces often both in my journals and in Emily's emails because, I think, it most accurately describes the sensation we had of riding the thought-waves focused on Simon and his healing at that time. It was as real and as concrete as a smooth stone placed in the palm of my hand and, although its impact is not as clear-cut as the effect of the brilliant intensive care he received, it too made a difference.

I also believe, with the nonscientist's reliance on faith, not proof, that if there is any type of brain designed to cope with massive trauma, it is Simon's. A highly associative thinker, an improviser, Simon had been adept at creating reality out of chaos in his solo improvisational work, and this to me was precisely the work required of him to resurface from the coma: a massive effort to re-create reality out of chaos. Was it Simon, then—his particular brain—that was the critical difference?

As a family we made a commitment to stay at Simon's bedside around the clock, to never leave him alone, and it is tempting to pride ourselves on making a difference. Was it because Emily was brave enough to sing him through the darkest time? Or because Marc had spent a lifetime sharing songs with Simon, and that shared musical experience formed a bridge that allowed Simon to return from the depths of his injury? Was it because Lorna, despite Si's annoyance, worked so tirelessly to open his mental files? Or because, at a more basic level, Simon inherited her DNA fire and because, like her, he pushes and strives, always? Or was it because, as Simon claims, he returned for me and Eli, that it was love that called him home? Did these things make a difference? Yes and no.

Yes, because when you reach that invisible dividing line between life and death, it is clear that each one of those individual moments matters. It all matters. And I am proud of us all; the generous sharing of human strength, skill, and spirit that I witnessed is, to me, nothing short of miraculous.

And no, because it would be a mistake to say that any one of our individual efforts might have made a difference. The brain is a mysterious organ, and each injury is unique. A neuropsychologist once told Simon and me that in her practice she had seen amazing things: people who had had horrific accidents— falling out of a helicopter, for example—who made significant recoveries and others who slipped on ice and were never the same again. Head injuries are impossible to predict, she told us. The brain is more than a sum of its parts. And I know that there is a mother today who sits, as I write this, at the bedside of her child, a child equally as gifted and bright and strong as Simon, waiting. A mother who pours just as much love as we did from the vessel of her body into the broken vessel of her child, who waits still. A mother looking for the key to wake her sleeping beauty, a key that is perhaps impossible to find.

There was no one key to Simon's recovery. It was and continues to be, like the brain, more than the sum of its parts. There was the miracle of intensive care units and soccer coaches and best friends and strong families. There was the miracle of music. And, then, we also got lucky.

AND TO ALL those who wait today, my thoughts go to them with the hope that amidst the deep waters of grief and fear they are also blessed with the buoyancy of great love:

Be strong.
Be safe.
These are moments in time.

{ 20 }

SHHHUGAR

ALTHOUGH IT IS only a ten-minute drive, the trip from VGH is exhausting for Simon. Eli and I travel with him in the GF Strong transport van; Simon, in an electric wheelchair, is tilted back at an angle to alleviate the sickening vertigo and nausea he experiences when upright. But even when he is tilted back, the movement and jostling of the van are excruciating. He doesn't complain, but Eli and I clearly see the waves of nausea rippling across his face with every bump and swerve. Arriving at GF Strong, he learns that, for the first time since being hospitalized, he will share a room with someone else. Richard, the smiling nurse who greets us, introduces Simon's new roommate, an elderly man named Wilhelm who suffered a stroke while cleaning his gutters at home.

The two hospital beds in Simon's room are separated by a thin curtain, and we are all relieved when Richard directs us to the farthest bed, the one adjacent to the large bank of windows that looks out onto the tree-lined residential streets surrounding GF Strong and not, thankfully, the bed closest to the door and the brightly lit, busy ward hallway. There is a shared mirrored sink on the wall opposite the beds and a door leading to a large bathroom, as well as closet space for both Simon and Wilhelm. Richard chats with Simon as the rest of us unpack the surprising amount of stuff we have accumulated over Simon's time at VGH. Lorna places the jasmine plant on the windowsill and hangs the few pieces of clothing we have purchased for Simon in the closet. Marc stores the boom box, the CDs, and a box full of letters on a shelf underneath the bank of windows as Eli and I hang the many photos of family and friends, along with the hand-drawn pictures from Simon's niece and nephew, Alice and Oscar, on a large corkboard. Richard and Simon continue to talk, their conversation punctuated by the exclamation mark of Richard's laughter. Richard is a tall man, and his playful manner is a welcome relief from the daily parade of serious and formal medical professionals we encountered at VGH. He is warm and generous, and his welcoming cheer is infectious. In short order we all feel more settled in. All the movement, activity, and conversation takes a toll on Simon, though, and as soon as he is airlifted into his new bed, he falls into a deep sleep.

MOVING DAYS ARE hard. This has been our experience with every ward or room change that Simon made at VGH. Each move is, however, a forward, positive one; we know that and are appreciative that this is not a move, say, back to the ICU. Still, the chaos of a moving day is difficult to navigate. Each new environment dramatically underscores how physically vulnerable Simon is. Each new environment has its personalities, rhythms, routines, and rules; having to constantly readapt is draining for us all. Things change in only a few seconds, the time it takes to drop ten feet through the air, and they keep changing and changing. It will never stop. On moving days, I know, as an empirical fact, that I will never have enough energy to keep up with the whims and caprices of constant change. The best I can do is hold on. Tomorrow, possibly, will be better.

GF Strong is a busy place. People in wheelchairs—some manual, some electric, some equipped with ventilator machines and controlled by a person's breath—roll through the hallways. People with crutches and braces on their legs propel themselves with a barely contained momentum down the hallway, while others shuffle aimlessly, poking their head into the nurses' station, the kitchenette, or other patients' rooms, as if searching for something vague but precious they have lost. Wilhelm, Simon's roommate, inches down the hallways with his walker. He barely acknowledges our presence, entranced as he is by the running water that flows from the tap into the sink at the far end of the room. He turns the tap on and watches the water for almost twenty minutes at a time, murmuring softly to himself.

While Simon sleeps, Lorna explores and returns convinced that Simon is on the wrong floor. The Spinal Cord Injury floor just below us is where he should be. It is, she argues, busier, brighter, and more cheerful. Not as dismal and depressing as the Acquired Brain Injury ward. When Simon wakes, she encourages him to discuss how sad he must be to find himself in this place.

"It's depressing here, don't you think?" she asks. "I find it depressing."

Once again, her question makes me angry and weary. Right now Simon isn't sad or depressed. He is less aware of his larger surroundings and more focused on the immediate issues of navigating this new environment. He is nervous but combatting anxiety by asking questions and trying to get a handle on what is expected of him starting tomorrow. Rehab means gaining independence, but what exactly does that entail? Does the staff realize that having a bowel movement is not an event he

can control? That he needs to travel everywhere with the catheter bag? What will happen in the morning? Does he have to
get up into the wheelchair for breakfast? Where will he go for
physiotherapy? For occupational therapy? Is he supposed to go
alone to these rehab sessions, or can I come with him? I am not
allowed to stay overnight in his room, as I have done at VGH,
so he is facing his first night alone. All these questions weigh
heavily, and he places a lot of pressure on himself to get things
"right."

Lorna is concerned about the appropriateness of Simon's
emotional responses. From her perspective, sad, depressed,
and anxious are the only intelligent responses to his new situation. I resent the suggestion that a downward spiraling of
emotions should be positively associated with his overall cognitive recovery, that somehow sad, depressed, and anxious
means he doesn't really belong here, on this ward of the walking or wheeling wounded. I ask her to join me in the hall so that
we can talk privately.

I talk; she listens. She talks; I listen. We don't come to a consensus except to agree that the tension of moving day is getting
to us both. We leave it at that. Lorna and I have very different
approaches, but we are both stubborn and opinioned and, on
this occasion, as exhaustion takes over, we reach a stalemate.
Also, we are both right. Yes, of course, Simon needs our support to feel contained, protected, and safe, but he also needs
the space to express the havoc of grief, anger, and fear that
must be brewing inside him. And as much as I fuss about the
need to cultivate a positive approach toward rehab, the truth is
I am terrified to open a door onto these more difficult emotions,
each one its own dark chasm, too deep and wide, too potentially bottomless, for us to attempt to cross right now.

Simon is still eating soft and pureed foods and has been
designated at risk for choking, so he is not allowed to eat in

the cafeteria on the main floor. Instead, we are directed to a small nurse-supervised room down the hall from his bedroom. Because the room is small, entire families are not allowed to eat with patients; I remain with Simon while Marc, Eli, and Lorna eat dinner downstairs.

It is a tough room, filled with some of the most severely injured or ill patients at GF Strong. A middle-aged woman sitting near the door steadfastly refuses the small bites of applesauce a nurse holds to her lips.

"Just one bite, Ms. Prinze," the nurse repeats over and over. "Just one bite."

In the middle of the room, another gray-haired woman feeds a young man—her son?—wiping his lips as he struggles to keep the food inside his mouth. Simon and I take a spot in the far back corner, angling his chair so that he can look out the window at the green tops of the trees lining Laurel Street. I open the foil lids of his milk, juice, and pudding cups and move the bundle of cutlery within reach of his right hand.

"It's a nightmare," he says, angling his head in my general direction. I am on his left, so it is difficult for him to face me directly. "It's a nightmare not having my legs."

I know, I want to say, but it's going to be okay. You're going to be okay. We're going to be okay. But I don't say it. It is a foolish sentiment, and potentially untrue. Perhaps even cruel. How can I undermine the weight of what Simon is expressing with pat, pacifying assurances? Everything happens for a reason. It will all work out in the end. In every cloud there is a silver lining. All the little fibs we appease ourselves with to simply not have to face a moment like this. But the potential of Simon's pain scares me more than anything I have ever faced in my life. I fear it will crack us both open. All the King's horses and all the King's men couldn't put Humpty together again. Can we?

"It is a nightmare," I say, swallowing a mouthful of tears. "I wish there was more I could do."

Simon smiles, the same smile he gave me as we wheeled away from his hearing test. It is the second time today that smile has saved me and, even with the left side all droopy, it is gorgeous. Complete. "I'll tell you what," he says. "Why don't you give me your pudding?"

BACK IN SIMON'S room, the nurses once again prepare him to be airlifted into bed. It is a complicated procedure, maneuvering the sling under Simon's body while he is seated in the chair. The sling must be aligned properly so that when he is lifted he doesn't dip or tilt dangerously.

Marc and Lorna say goodnight, and as I walk them to the elevator, Lorna sighs. "He's like a baby, in that sling," she says, "a baby in the mouth of a giant stork." It is a strikingly apt image, one that conveys the continuing horror of his helplessness, a horror that only deepens the further away he is from the coma days.

When I return, I pull the sheets back from Simon's feet to check his bandages. He left the ICU with large pressure sores on the tops of his ankles; they are mostly healed now, though the still-delicate skin is discolored and scarred.

"Your toenails are really long, Dad," Eli says from his seat on the window ledge. "Can I cut them for you?"

We exchange positions, Eli picking up the small bag of toiletries he and I assembled for Simon at the London Drugs across the street from our hotel. I pull my pen and notebook out but only pretend to update my journal, watching as Eli gently manipulates Simon's swollen, unfeeling feet. He takes tentative little snips on the overgrown rind of the big toe.

I was terrified of clipping Eli's nails when he was a baby. I let his fingernails grow so long he scratched his face, and when I did finally cut them I nicked the skin of his newborn middle finger, releasing a bright red bubble of blood. This, I fretted, was motherhood: despite my best intentions and best attempts,

I was doomed to cause harm to my child. I insisted on deferring the task of nail cutting to Simon, with his steady hands and steadier nerves. I watch now as Eli returns the gesture while Simon inquires about his new job at the marina.

"I'm a glorified bellboy," Eli says as he pulls the emery board out of the toiletry bag and begins filing his carefully clipped corners. "Only not so glorified."

Since Simon's accident, Eli has adopted a new look, buzz-cutting his hair and buying a UFC TAP OUT shirt, as if these trappings of the hypermasculine, the super-strong, can protect him. Still, the gentleness he inherited from his father flows through his every movement. After he finishes filing the nails, he buffs away the dead and calloused skin. We are okay, I think, as Eli squeezes a pearl of peppermint cream into the palm of his hand and slowly works it into the dry skin of Simon's feet. We are okay right now. This time it does not feel like a thin lie, a rope bridge flung across an abyss; it is real and true. We are okay right now. This is all that matters.

Simon falls asleep. I leave the cell phone tucked into the side of his bed and a note attached to the bedrail—*Sweet dreams shhhugar; I'll see you at sunrise*—and Eli and I walk along Laurel back to Broadway and the hotel, the heat of the day dissipating into a cool blue dusk.

When I return in the morning with tea and muffins, Simon is already awake.

"I liked the note," he says. "Especially the 'see you at sunrise.' It was like a little prayer through the night. I had butterflies in my stomach waiting for you to come."

His face is flushed. I recognize the look: it is one of new love, of falling head over heels, and my face warms with a reciprocating glow. Simon and I are a few years shy of our twentieth anniversary. After so many years, and despite whatever traumatic events conspire in provoking it, that spark of fresh love,

that ache of anticipation, is delicious. There isn't much time
to savor it, though. Wilhelm is up and running the tap water at
the end of the room and then Sean, Simon's physiotherapist,
arrives with a breakdown of the day's schedule. After the nursing staff help Simon through his morning routine, we are to
meet him in the Occupational Therapy gym.

The OT gym is large and busy, filled with both gym equipment and items of "daily living"—a mattress, a kitchenette—as
well as work tables, where several people practice fine motor
skills: picking up pennies, doing up buttons, tying laces. Sean
waits for us by a raised exercise plinth, flanked by two students
who will assist and observe Simon's session. Sean is stocky
and strong-looking but a good deal shorter than Simon, and I
watch with barely contained anxiety as he prepares Simon for
something he calls a sliding board transfer. Angling Simon's
chair next to the plinth, Sean places one end of a smooth, oval-
shaped board under Si and the other end on the cushioned mat
so that the body of the board bridges the gap. Simon leans his
full body weight forward into Sean's shoulder while one student steadies the board and the other, holding onto the fabric
of Simon's sweatpants, lifts from behind. In one smooth movement, Sean swoops Simon off the chair, across the board, onto
the plinth, and down onto his back.

"And that's how a sliding board transfer is done," Sean says.
He and the students then proceed to move all of Simon's limbs,
asking Simon to flex or extend, to lift or bend. The strength
of each muscle movement is graded on a scale of 1 to 5—the
Oxford scale, 0 being no muscle movement and 5 being full
strength. Below his waist, Simon's scores are all 0; the right
hand and arm, for a variety of movements, scores 3s and 4s;
the left scores 1s and 2s. Simon is unusually unchatty, focusing all his attention on Sean's directions. The focus pays off:
with Sean positioning his arm, Simon is able to raise it almost

above his head. Despite the low left hand and arm scores, I am impressed by a whole new level of left arm movement I haven't yet seen. It figures that Simon would save his best performance until there is an audience.

Sean accompanies us back to Si's room and attaches rubber TheraBands to the headboard and side rails of his bed, instructing Simon in a few strength-building exercises.

"So you can keep working over the weekend. Get you ready for Monday," Sean says, "when the real work begins."

"What did you think?" I ask Si after Sean leaves. After the small physio room at VGH, the size and scope of the OT gym is impressive, and I like Sean: he is both gruff and friendly, and I thought he did a good job at establishing an immediate level of trust with Simon. But I'm not sure how it would feel to be the person lying on the plinth being poked and prodded and measured.

"Sean is very meat and potatoes," Simon says, lids heavy with the pressing need for sleep. "That's what I need: meat and po—" He trails off midsentence into a deep, impenetrable, well-deserved slumber.

{ 21 }

OUTSTANDING

WE ARE ALONE the next day, or at least as alone as you can
be in a massive rehabilitation center. Wilhelm's wife
picked him up Friday night so that he can spend the
weekend at home. Lorna stays at Jer and Barb's, taking a much-
needed day of rest—her first since Si's accident—and Marc is
taking Eli to his annual soccer tournament in Chilliwack, along
with my brother, his family, and my mother. Simon insists that
we should be in Chilliwack too.

"We can't miss the tournament," he says over breakfast.
"Let's just go. We'll come back here when it's over."

I understand the sentiment perfectly. There are so many
times when I have a similar impulse: Simon and I just need a
quick time-out from this hospital scene. A small break to take
in a soccer game or go for a walk in the woods or snuggle on
the couch, just to recharge. Then we'll get back to all this. But
I also see clearly now that Simon has not fully acknowledged
the extent of his injuries. I explain in detail why driving to

Chilliwack and staying for the tournament isn't possible yet and end by making a promise: "Next year. We'll go next year."

Simon nods. It is half an hour before game time, and he asks me to dial Eli's cell phone so that they can speak. They chat briefly about the lineup of games for the day and then Eli says he needs to go warm up.

"Good luck. I love you, man," Simon says. He passes the phone back to me, his face betraying a kaleidoscope of emotions that resolves into sudden tears. It is shocking. In all our time together, I have never seen him cry. I have seen him tear up when his friend James died and when our dog Ananda was hit by a car, but I have never seen him cry.

"I am a full-grown man, but I'm living like a baby," he says. "I hate feeling like a fucking baby."

Hand pressed to face, his shoulders shudder as he gulps for breath. I sit beside him, paralyzed by my own helplessness.

"I'm sorry," he says finally. "I'm so, so sorry. I feel better for crying." He dries his tears with his good right hand and returns to work on the left, pulling and pumping away on the TheraBands.

A SCHEDULE IS quickly established at GF Strong. Mornings, Simon works in the physio gym with Sean; afternoons, with an occupational therapist, Melissa, in the OT gym. Simon will see various other professionals during his stay at GF Strong—his rehab doctor, a social worker, a neuropsychologist, a music therapist, a sexual health therapist—but the relationships that develop and deepen most quickly are with Richard (King Richard, as Simon and I come to call him), the nurse who initially greeted us at GF Strong, and with Sean. It had been challenging, Sean confides to me during the first week, to figure out a plan for Simon's physical rehab program. Simon has three distinct injuries: a brain injury, a brain stem injury, and a spinal cord injury. Sean explains that, separately, he would approach each one of those injuries in a very different manner. It seems, however, that both Sean and Simon are up for the challenge: that first full week at GF Strong, there is a quantum leap forward in Simon's ability to move. On September first, ten days after the move into rehab I write in my journal:

I can't remember the exact day you started lifting your left arm independently. When we left VGH, you couldn't lift it. Now you are holding it straight out for Sean. You are holding a teacup, cutting meat with your knife. Amazing.

THERE IS, HOWEVER, a new development along with Simon's increased activity: the onset of pain. This pain is not like the previous aches or soreness Simon experienced in his arms and neck at VGH. This is something entirely new, the mysterious and elusive neuropathic pain that often accompanies spinal cord injuries and usually starts a few months after the injury has occurred.

"It's like random bolts of lightning in my bikini line," Simon explains to Sean and me.

"Sounds like my college years," Sean says, and we laugh. But it isn't funny. At night, Simon is often woken from a deep sleep crying because of the pain that surges through his hips and lower back. It is cruel, the irony of an injury that robs one of sensation and movement, of everything but pain.

AT THE BEGINNING of September, Simon has his first team meeting. His parents and I are present, as are Sean; Richard; the rehab doctor, Dr. Yao; the occupational therapist, Melissa; Mary, Simon's social worker; and Janet, a representative from WorkSafeBC. We gather in a generic conference room, the unadorned whitewashed walls and uncomfortable chairs an ode to parsimonious practicality. The only dash of color is provided by Dr. Yao's stylish turquoise pumps and matching handbag. The elegant cut of her skirt and blouse momentarily distract me, and I wonder how many days I have worn the same pair of jeans and crumpled T-shirt. Too many. A sigh escapes me as I sit down beside Simon, crossing my legs and settling my sneakered feet underneath me. We position ourselves in a horseshoe around Simon; introductions are made, roles are outlined, and Simon's current health status is summarized. Next the team wants to know: What are Simon's rehab goals?

"Aside from walking out of here?" Simon says from his tilted-back position in his electric wheelchair. "I want to see Eli play a soccer game. I want to play guitar. I want to get back home. Be a husband; be a father."

After the meeting I meet separately with Janet, who is to become Simon's caseworker in the Special Services division of WorkSafeBC. Janet is tall, dark-haired, very professional but also friendly. She congratulates Simon on his amazing progress in rehab, and they make a good connection. Still, I am wary heading into the meeting; I have no idea what to expect. Will I have to argue and fight for Simon to receive benefits? I have had no firsthand interactions with WorkSafeBC—or any insurance company, for that matter—but experience has taught me that large institutional bureaucracies tend to be less than helpful to the clients they are meant to serve.

It is a weird and one-sided meeting. I ask a few questions, but for the most part, Janet outlines what WorkSafeBC will and will not do. Previously, injured workers had the option of receiving either a single lump sum settlement or a monthly payout until the age of sixty-five. Now the only option is a monthly payout. Medically necessary procedures and equipment are covered, and there are allowances in a variety of areas: clothing, exercise equipment, recreation, and vocational retraining. Necessary renovations to make housing accessible are covered. The amount of the monthly payments is determined by the level of permanent injury, and the maximum payout is 90 percent of one's pre-accident salary.

Simon's sense of smell is gone. His right ear is gone. A piece of his temporal lobe, gone. His legs. Gone. I hate the word *permanent*. I start to shake.

"I do not want to make any predictions about what Simon is capable of doing in rehab, but, given the severity of his injuries, he will most likely qualify for the full 90 percent." Janet smiles, kindly. "Several times over, actually—with the loss of his legs, the loss of sexual function. It is a long process before all this is actually confirmed, so nothing is guaranteed, but I want to give you as accurate a picture as I can."

Simon's parents are relieved when I return and outline what Janet has told me. Part of me is relieved too. Ninety percent of Simon's current salary is a workable amount of money. Significantly less, yes, without Simon's extra gig money and my income, but I can run the household on that amount, at least temporarily. But the relief I feel is distant, abstract, a pale beacon in the cold fog that has filled me since my talk with Janet. I don't blame her. She handled a difficult situation with sensitivity. But it is horrible. Awful. This breaking up of Simon's body parts into an income percentage, a morbid living dissection. Our current financial relief bought by his ability to take a piss, have a dump, get an erection. Has Simon lost his sexual function? What is sexual function or pleasure without sensation? What would that mean to him? To me? To us? These are questions too big and messy for either of us to face at the moment, so I torment myself with the idea that each body part is worth a certain amount of money. What would you pay for a right ear? A left hand? A penis with the ability, once more, to ejaculate?

Many would categorize SCI as one of the greatest survivable catastrophes experienced by a human being.

—RICK HANSEN INSTITUTE

AS SIMON ENTERS into his rehab program, the focus for him is on his spinal cord injury. Once again there is a very sharp learning curve as we gather and process information. This includes learning the acronyms of spinal cord injury, or SCI. An injury is classified according to two related measures: whether it is a complete or an incomplete injury and where it ranks on the American Spinal Injury Association, or ASIA, scale.

The spinal cord is the multilane pathway that connects the central nervous system with the rest of the body, channeling the world of sensation from our feet and fingertips to our brain, and the brain's subsequent commands back to the feet and fingertips, in a continual loop. Bony vertebrae surround the spinal cord and provide a sturdy shelter. The spinal cord itself is continuous with the brain stem and, like the brain, is surrounded by three layers of protective tissue: the dura mater, arachnoid mater, and pia mater. In opposition to the organization of the brain, the spinal cord has an inner core of gray matter and an exterior composed of myelinated white matter. It is the level of damage sustained by the spinal cord—not by the vertebrae— that indicates the level of injury. Most often, damage to the spinal cord occurs when it is squeezed, pinched, or bruised, injuries that are further complicated by swelling or a decrease in blood flow to the injured area. This type of spinal cord trauma constitutes an incomplete injury, and although initially these injuries often present as full paralysis, there is the potential—especially with the dramatic improvement over the last twenty years in acute care and rehab for spinal cord injuries— that some recovery of motor function and/or sensation can occur. A smaller percentage of spinal cord injuries occur when the cord is entirely, or almost entirely, severed, as is the case for Simon. This type of injury is classified as a complete injury.

An ASIA A, or complete injury, means there is no movement or sensation below the level of injury. In an ASIA B injury, there is some sensation but no motor function. For an ASIA C, there is both sensation and motor function below the injury, and this only increases for an ASIA D. The farther into the alphabet you are, the more hope there is for some recovery while a complete injury equals almost no hope for any significant change to occur. Generally, rehab for a complete injury is geared not toward healing the injury but toward learning to live with it.

Simon was classified as an ASIA A, complete injury, at the time of his spinal cord operation, but because he was in a coma, none of the typical exams could be performed. Now he has whispers of sensation in his right hip flexor, which is, technically, below his level of injury.

"The doctor thinks it might be possible I'm an ASIA B," Simon tells Sean.

"You're an ASIA A floppy para all the way," Sean says. "Sorry, man."

"Fuck," Simon says, smiling. "You know I never wanted to get a D so badly in my life."

Both Simon and I rely on Sean's no-bullshit approach. The reality is that, with the head injury complicating his recovery, Simon will have difficulty reaching even the basic level of mobility and independence that someone with only his level of spinal cord injury could achieve. It also becomes apparent as Simon and I meet his fellow GF rehabbers that a classification of an incomplete injury can be a mixed blessing. Often when people with an incomplete injury are up and walking they are, to the world at large, perceived as "healed." But for many of them some of the more devastating effects of a spinal cord injury—the bowel, bladder, and sexual dysfunction, the debilitating neuropathic pain, and the fatigue—remain but are invisible to the outside world. Often, for those with a diagnosis of an incomplete spinal cord injury, hope traps them in an uncertain, seemingly unending, and extremely challenging rehab regime. In the cafeteria one day, we meet a young man, a mid thoracic complete injury paraplegic, who sums up the contradiction beautifully.

"When I arrived here, there were other patients on the spine ward in electric chairs with no movement at all and I felt like I was so much better off. Three months later, some of them are walking out of here. It's hard not to feel a little jealous. But in a lot of ways, I'm glad I'm a complete injury. There's nothing I can do about that," he says, "which means I can focus on the things I *can* do something about."

Still, Simon cannot quite yet accept the prospect of no hope. Although I have clearly outlined the doctors' diagnosis of a complete spinal cord injury, it is inconceivable to him that his legs will not heal. And there is no shortage of well-intentioned but ultimately misguided visitors who want to share the inspiring story of a friend of a friend who was told he would never walk again but who, three months later—three months!—is running marathons. When Simon asks me, as he does randomly and often, what I think will happen with his legs, I don't know what to say. When Simon was in the ICU, I reckoned with the fact that there would be little room for improvement with the spinal cord injury. Legs schmegs, I said, just give us his brain back, and I meant it. But I also believe that how healing is approached can potentially affect the outcome. And—maybe— I'm hoping too, blindly and unreasonably, that there is still a little more injury-defying magic left in Simon's body.

So I can't say "Simon, you will walk—I know you can." But, even after reminding him of the doctors' predictions, I also can't say "No, you will never walk again." I say the only honest thing I can: I don't know. I don't know what your body is capable of. I don't know what the future of medical research will bring. I don't know. But for now, let's commit to continued healing.

LORNA AND I return to the coast to go house hunting. I am haunted by Simon's absence. For the past three years, buying a house has been a central preoccupation for us. We saved a down payment nest egg in our retirement accounts, but in the prohibitively expensive West Coast market, our meager savings couldn't buy much. We looked seriously at one place in the springtime: a beautiful heritage home on the outside, a structural mess inside. With that last disappointment, Simon and I decided our best course of action was to buy and pay off a piece of land while remaining in our wonderful little Hobbit house with its exceptionally reasonable rent and then, eventually, take out a loan to build our dream house. For his birthday, I bought him a book on innovative architecture that incorporated alternative, off-the-grid energy sources. We spent cold spring evenings by the fire pawing through the pages, trying to better understand the difference between passive and active energy systems and discussing the feasibility of wind turbines, geothermal energy, solar panels.

Still, even without Simon, it is a pleasure to have a day away from GF Strong imagining a potential future in the various houses Lorna and I look at. With the promise of financial support from Simon's family, Lorna and I create a list of houses, all bigger and better than anything Simon and I viewed before. There are several promising options, but it is Lorna's last-minute addition to the list that we fall in love with. Newly built, the house has a beautiful wood finish and gorgeous windows, an open concept that incorporates a raised hull-shaped ceiling, an intelligent, streamlined design with magnificent built-in shelves and cupboards. Even though it is at least a forty-minute drive away from our Halfmoon Bay neighborhood, it is simple and elegant, a gazillion times nicer than any other house we see that day. I worry that it is too expensive but, swayed by Lorna's insistent enthusiasm, put an offer in that evening.

The next day, Simon and I travel by transport van out to UBC for a CT scan. Everything goes smoothly until it is time to return to GF Strong. The driver who arrives to pick up Simon is cross and miserable, loudly complaining about his wife, children, and grandchildren as he tries, and fails, to lock Simon's chair in properly. I sit in the front seat listening. As the driver bitches and Simon makes polite and helpful suggestions, a sudden and unexpected spasm, a hailstorm of fury, crystalline and pure, rolls through me. I hate this complaining man. I hate WorkSafeBC for performing a cost analysis of Simon's injuries. I hate the thought of buying a house without his seeing it. I hate the new house for being something we could never have previously afforded. I hate the new house for what it has cost Simon. For what it has cost all of us. As the driver leaves to retrieve a second passenger, I start to cry.

It is the first time since his accident that I have cried in front of Simon, and it upsets him. I bend down, apologizing, and properly secure the wheelchair locks, for the moment beyond exhaustion and unable to stop the landslide of tears.

The next day Wilhelm is moved to a new room, and Simon's new roommate, a seventeen-year-old boy named Danny, arrives. Neither Simon nor I have a chance to say good-bye to Wilhelm, but he appears later that day, silent as ever, to execute his regular evening routine in this more familiar room. As he turns on the tap of the sink at the foot of the beds and watches the water run, my cell phone rings. It is the realtor; the sellers have countered my offer with a new asking price. I decline it and, although Lorna and I return a few times more to view it, we make no further offers on that beautiful house.

I RECOGNIZE DANNY, Simon's new roommate, from the hallways of the spinal cord ward at VGH. I saw him dressed in the same white TED stockings and blue shorts that Simon wore there, being wheeled around by his family. Mother, father, a younger brother, no more than fourteen, glued to the armrest of the wheelchair where his big brother's hand lay motionless.

Danny is a year older than Eli. Seventeen. His spinal cord injury is a high one, and he does not have the use of his hands or arms. He is tall and impossibly thin after his hospital stay, and his golden mane of hair looks to be the heaviest part of him. Well-spoken and poised, especially given his current situation, he is a fair-skinned beauty who I am certain sets teenage girls' hearts aflutter. His parents and younger brother visit, along with his dog, but they live out of town and can't be there every day. Unlike many of the people at GF Strong, Danny has no insurance claim helping to defray the cost. His mom and dad have to work, and in their absence, and Eli's, Simon and I, parents to the core, worry about him—secretly, when he is out of the room at his own physio sessions.

"Look at that," Simon says, pointing to the mini-fridge that Danny's friends have left on his side of the room. It is stocked with Snapple and chocolate puddings. "He eats like a hummingbird. He'll never gain any weight back."

"His mom brought him homemade soup yesterday," I say. "That's exactly what he needs. Something warm."

Danny has a hard time falling asleep at night and a hard time waking up in the morning. The nurses who feed him breakfast in the morning have a tendency to scold him about this, to transform from professional health care providers into overworked and vexed mothers as they ladle runny porridge into his unwilling mouth. He, in turn, is sometimes argumentative with them. On the other side of the curtain, Simon and I listen, our hearts breaking. We know this young man. We know this boy. He could be Eli or any one of his friends. We have made him pancakes on Saturday morning, driven him home wet and tired from soccer practice, picked him up from that party—you know that party, the one where the parents remained behind closed doors on the second floor and everyone had their first beer. Turned the music up a little too loud. Danced with something resembling your soul at stake. Stole sweaty kisses in dark corners. Came home, every atom still dancing, jubilant, knowing you would never be tired again. That party.

DURING THE FIRST week of September, I sit down with Simon's rehab doctor, Dr. Yao, to discuss the results of Simon's latest CT scan.

Extensive encephalomalacic changes are present in both frontal lobes, worse on the left than on the right, which has progressed since the prior study with resultant sulcal widening. Post-traumatic tissue loss in the middle cranial fossa is present and has evolved in appearance in an expected manner.

"Tissue loss, extensive enceph-al-o-mal-a-cic changes," I read out loud, stumbling over the excess of vowels. "That's not good, right?" By now I understand that it is impossible for damaged or destroyed neurons to heal or regenerate. The injured or absent areas will remain that way.

"No, it's not good," Dr. Yao says. "But clinically, day to day, Simon is improving, and that is good. That is where we will take our cues."

"He's sleepy right now, but, you know, he doesn't feel any gaps in his thinking."

"Well, that," Dr. Yao answers predictably, "is symptomatic of a brain injury. It is common not to feel like there is anything wrong."

I don't see any major gaps in his thinking. This is what I want to say but don't. And now that Simon is less confined to bed and awake more, it is no longer entirely true. There are differences. The focus he requires to make his therapy session appointments precludes any other form of interaction, and if someone tries to engage him briefly in conversation, he is often preoccupied and impatient, impulsively driving off in his electric wheelchair midsentence. He likes to drive the chair fast, the New Driver sign dangling precariously off the headrest, and his solution to hallway traffic jams is generally to speed up and barge through. He worries incessantly about each new step he takes with Sean and Melissa, and it is difficult, if not impossible, for him to take a mental break. He often has problems with the proper sequencing of events: if ten small steps are needed to solve a given problem, he might start on step one but then leap ahead to nine or ten right away. There are occasional and surprising gaps in the logic of his everyday thinking, more evident at the end of the day when he is tired.

"Before his accident Simon was extremely bright and quick...," I say, letting my sentence trail off, unsure of how to frame my question. Unsure, exactly, what my question is. "He's in the OT room, picking up pennies, and his hand is getting better but how..." I stumble again, afraid that the hot tears, always so close to the surface, will boil over. Deep breath. I focus on Dr. Yao's shoes: ruby red with a blocky, tapered heel. As always, they match her purse, and, as always, I find them elegant, enviable, and grounding. Another deep breath. "How do you heal executive functions?" I ask. "How do you heal thinking?"

"There are no blueprints," Dr. Yao says. "The brain's ability to heal is contingent on its ability to make new connections."

This makes sense and is surprisingly comforting. Simon's great gift, as a musical improviser, is his ability to make spontaneous and often startling new connections over the course of just a five-minute song. So, I tell myself, despite the mushy frontal lobes and the missing bits of the left temporal lobe, there is the possibility of continued healing. For his thinking, at least, if not for his right ear. With Simon defying so many of the doctors' predictions, I have been holding out for the possibility of healing in his right ear. This CT scan shows, however, that one of the many fractures—through occipital, temporal, and sphenoid bones—starts at the back of Simon's skull, near the top of the spinal cord, and traverses through the temporal bone, across the canal of the carotid artery, and through the semicircular canal of the inner ear and the cochlea, ending close to the Eustachian tube. There is no need, Dr. Yao asserts, for further exploration. Damage to the structures of the inner ear does indeed mean the hearing loss is permanent.

ARMED WITH THESE CT results, Simon and I take the GF Strong transport bus back to VGH for an afternoon appointment with Dr. Haw. I saw Dr. Haw only once in the early days of ICU. He was surrounded by students and other doctors engaged in their early-morning rounds, and we had only the briefest exchange before he pressed on. Simon, of course, has no recollection of him, but this is the man who has been inside Simon's head. Dr. Haw has seen the folds and fissures of Simon's cerebral cortex, touched the living tissue that has, throughout his life, defined his particular Simon-ness.

"It is a pleasure to meet you, sir," Simon says, extending his right hand.

"And a pleasure to meet you." Dr. Haw grips Simon's hand. "When we first were introduced, you were more than halfway through death's door. I'm glad you've returned."

"Me too," Simon says. "Thank you."

"Thank you," I echo, acutely aware of the feebleness of my language. Only an hour before, I said thank you to the woman at Starbucks when she handed me an Earl Grey tea. Thank you is not enough to offer Dr. Haw, but at the moment, it is all I have.

The room is small, the narrow space between Dr. Haw's desk and the wall behind us underscoring how cumbersome the electric wheelchair is. I push chairs into the far back corner so that Simon can squeeze in as Dr. Haw settles himself behind his desk. He is a short man with a kind and serious face. The office walls are hung with various degrees bearing his full name—Dr. Charles Sung Haw—and I am charmed by the verbal suggestion of musicality, the hint of humor, in his name; I imagine I can see this in his face, too.

Dr. Haw asks a few questions and patiently answers ours, and then we discuss when the large section of skull bone still missing from the left side of Simon's head will be replaced. Dr. Haw wants to replace the bone flap as soon as possible, but two things need to happen before he can proceed: the open and miserable pressure sore on the back of Simon's head will have to heal entirely, and one of the drugs Simon is currently on—heparin, a blood thinner used to prevent the blood clots that spinal cord injury patients are susceptible to in the first few months after their injuries—will have to be completely out of his system.

"Perhaps we'll be ready for surgery in four to six weeks," Dr. Haw concludes as he walks us to the door.

"Thank you," Simon says again and, again, I echo it. *Thank you.*

A week later we return to VGH, this time for an appointment with Dr. Dvorak, the surgeon who operated on Simon's back. I have not met Dr. Dvorak, only his assistant, Dr. King. At the time of Simon's spinal cord surgery, the head ICU doctor assured me Dr. Dvorak was, if not the best, one of the best spinal cord surgeons in the country.

We follow a receptionist to an examination room where we wait, busying ourselves by studying the large diagram of the spinal cord hanging on the back wall. Soon we are joined by a doctor who looks too young to have achieved the status of best spinal cord surgeon in the country.

"I'm Dr. Cabot," he says, shaking Simon's hand. He turns and shakes my hand. "I don't know if you remember me? I was the doctor in the ICU the first few nights Simon was there. I remember you both and"—he turns back to Simon—"it is just outstanding to meet you now. You've come a long way since that time."

He explains that he will do a brief physical exam, and then Dr. Dvorak will be by to discuss the results of Simon's most recent X-rays. We discuss Simon's dizziness, which is still a significant issue, and he asks a few general questions before instructing Simon to raise and lower his arms and to flex and extend and squeeze his hands, much the same way that Sean did.

"Outstanding," Dr. Cabot repeats, over and over, as Simon executes each movement. "This is outstanding."

"My physio is helping me move to a manual chair," Simon says, taking the opportunity for a much-deserved brag. "Wheeling is hard for the left side, but he thinks I can build up the strength."

"It's outstanding. This—things like this—it's why I wanted to become a doctor," Dr. Cabot says, both palms open, gesturing toward Simon. "I'm going to get Dr. Dvorak now."

A few minutes later they both return. Dr. Dvorak has a friendly face and a largesse in his physical stature that seems to easily accommodate the mantle of one of Canada's top spinal cord surgeons.

"It's a pleasure," he says, shaking both our hands. He explains briefly the placement of the rods in Simon's spine. The most recent X-rays look good, and there are no new developments or problems. When Simon asks what kind of healing he can expect, Dr. Dvorak says that now, only time can tell.

"But you have certainly defied our previous expectations. You know, there was serious talk at a certain point when you were in ICU of not even doing the spinal cord surgery, the expectation that you would live being so low." Dr. Dvorak smiles a wide-open grin. "I'm certainly happy that we made the decision to proceed when we did."

"Me too," Simon says. Once again we both offer up our thin, insufficient thank-you's and agree to book follow-up appointments at Dr. Dvorak's new location at the soon-to-be-opened Blusson Spinal Cord Centre.

As they both exit the room, Dr. Dvorak turns to Dr. Cabot. "Holy cow," he says. "It's unbelievable."

"I like that Dr. Dvorak," Simon says as the door closes behind them. "He looks just like Bobby Orr."

I BUY A paperback copy of Norman Doidge's *The Brain That Changes Itself,* my heart quickening as I read the preface: "This book is about the revolutionary discovery that the human brain can change itself, as told through the stories of the scientists, doctors, and patients who have together brought about these astonishing transformations."

This book becomes my new bible. I carry it with me, returning to different, specific passages, over and over again, the pages quickly becoming soft and slippery.

Since the beginning of modern science, the dominant governing metaphor for brain function, Doidge explains, was one of a "glorious machine," and although machines were capable of many extraordinary feats, their core properties and capacities were fixed and unchanging. In this paradigm adult brains were "hardwired"; the various components of the brain were designed to perform specific functions; if the brain sustained any type of injury or damage, it was assumed to be permanent. Even a healthy brain, within this framework, could not be improved through intense physical or mental activity.

In his book, Doidge sets out to show how limiting and erroneous the machine metaphor is. Rather, the intricate neural

circuitry of the brain has the potential to be highly fluid, mal-
leable, constantly rearranging and reorganizing its pathways.
The brain—that creative sculptor of reality—has the capacity to
change its own structure and functioning through the thoughts
it thinks and the actions it takes. In making his argument for
the brain's "plastic" quality, Doidge relates several breathtaking stories: a man, blind since birth, who regains functional
vision through a device placed on his tongue; a young woman,
born with only one cerebral hemisphere, who is an active member of her family and community; a man cured of chronic pain
in his phantom limb by an ingenious and inexpensive mirror-
and-box invention. An elderly poet and teacher who, after a
massive stroke in his brain stem, recovers sufficiently to return
to his previous life, teaching, remarrying, and dying years later
while hiking in Bogatá. I read these stories over and over again,
trying to learn the language of neuroplasticity. I read these stories over and over and discover the exact thing that I had been
so afraid of during the time in ICU: Hope.

{ 22 }

IT'S A DRAG

FRIENDS OF DANNY'S parents were killed traveling from Nelson to Vancouver. The side of a mountain shook loose and landed on the stretch of road their car zipped along.

"*That* was an accident," Danny tells me. According to Danny, what happened to *him*—a missed landing in a bike jump—was not an accident. "It was stupid," he says. "There was this moment and I should have pulled the front wheel up. I knew I should pull it up. But I didn't. It's my fault."

I hear in his words that he lives this fraction of a second over and over again. In flight. Creating his own breeze in the hot blue air of a breathtakingly average summer day.

"Everyone freaked out when I landed," he tells me. "Even though I couldn't move, I had to talk them through calling my mom. Getting an ambulance."

Over and over again. The inhale. The moment of suspension. The brief but glorious tricking of gravity. The slight shifts in body weight. Forward. Backward. The rubber of the front tire, touching down. Exhale. Wrong. The twisting catastrophe that lies ahead.

Over and over. He practices it over and over in his mind. Given the chance, Danny would never miss a landing again.

"It was an accident," I tell him. "Everyone always thinks they could have done something to prevent a bad thing from happening. But that's why it's called an accident."

"Yeah," he says, and I almost see the phantom shrug of his still motionless shoulders. "I guess."

THE VERTIGO GRADUALLY eases, and eventually Simon is deemed no longer at high risk for choking and is cleared to eat the solid foods in the general cafeteria. In the area of bowel and bladder care, however, there is a frustrating lack of improvement.

"You must hate this," Simon says one morning, his tone surprisingly harsh. Accusing. I am helping him with his bowel routine, trying to get through it without nursing support. "You must hate having to do this."

"I hate that you have to go through it," I say. "But I wouldn't be anywhere else."

"I'm so sorry," Simon says, his voice softening. He says this often, randomly apologizing, and it always shocks me when he does. "Stan, I'm so sorry to do this to you."

"Don't apologize, Beau." It's too much: having to live through what he is living through *and* his need to apologize all the time. "Please don't be sorry. You have nothing to be sorry for."

JOHN, DOWN THE hall, brings Simon graphic novels. He often parks his electric wheelchair at our table in the cafeteria, joining us for lunch and a conversation that covers a wide range of topics, everything from the intimate to the philosophical. Discussions often include the trials and tribulations of rehab, the intricacies of interpersonal relationships, and the state of arts funding in Canada. Simon and John usually wind up by discussing their favorite sci-fi stories. Ursula K. LeGuin, they agree, is a genius. Isaac Asimov too. John is a veterinary surgeon from the Yukon, a man who, confined in the rehab center, confined in his injured body, is articulate and passionate in his longing to return to the wide open spaces of the North. A motorcycle accident on a late-night highway sent him here, to Vancouver, with a high-level incomplete injury and a hard knock to the head. The months he spent in the hospital before coming to GF Strong are lost to him, his brain a sieve through which days passed, leaving no silt trail behind.

"Even though I seemed to be functioning normally," he explains to us one day in the cafeteria, "I couldn't remember anything from day to day. People would visit, and a week later, when they returned, it was as if I had never seen them."

Pain is the ballast that eventually allows his days to accumulate once more. As his spinal shock wears off, sensation returns, but it is distorted, amplified, shocking, and relentless in its intensity, and he remembers: a doctor blowing on the skin of his forearm and igniting a wildfire in his nerve endings. An electric storm caused by the fall of a shirt sleeve or the brush of a bedsheet. Pain, a hard, bright light.

THE EFFORT THAT Simon puts into his rehab is epic. The physio gym, unlike the OT gym, is dedicated solely to moving and exercising the body. In the back corner are parallel bars for those who are able to practice walking. The rest of the L-shaped room is lined with dark blue plinths and littered with various props: hand weights, medicine balls, and TheraBands. Here, Simon first works on balancing. He sits on the edge of a plinth, and he and Sean (or I or Eli or Marc or Lorna) toss a large beach ball back and forth. Balancing is both extremely difficult and scary for Simon, but this work eventually enables him to sit and balance long enough to put on his own shoes. Next, Sean works on building up arm strength, urging Simon to four, five, and then six! seated push-ups on the plinth; this rebuilding of gross motor strength eventually allows Simon to accomplish sliding board transfers competently enough that he no longer needs to be flown, a giant baby in a stork's bag, back into bed. Sliding board transfers lead to transfers onto a raised toilet seat, then onto a shower bench, and eventually into the passenger seat of a car. The transfers into a car are particularly scary for me to watch or help Simon with as the effort it takes to make the movement requires him to pivot and swing his head with force. Simon is long in his torso, and several times the vulnerable left side of his head, still with no skull protecting the brain, comes precariously close to thwacking the frame of the car. But Simon trusts Sean and his meat-and-potatoes tough love attitude; Sean's approach keeps Simon pushing hard every day.

Si's occupational therapy sessions cover a broad range of activities that fall under the general term "life skills." At first the sessions focus on the fine motor skills of Simon's left hand. At the urging of either Melissa, the OT, or Natalia, the ever-cheerful physiotherapy assistant, Simon picks up pennies or fits small metal pegs into different-sized holes, activities reminiscent of games for toddlers. Simon's frustration and boredom are, however, mitigated by the effectiveness of these exertions. More and more mobility returns to his stiff left fingers. Next he works on applying the crude strength he is regaining in physio to daily activities like dressing and showering, activities once taken for granted that now have to be broken down into small steps for him to be safe and successful in accomplishing them. The OT sessions also focus on proprioception, or an awareness of one's body in space, a faculty that is affected by both head and spinal cord injuries; a lack of awareness can mean banging an unfeeling knee into a wall or catching and dragging a foot under a wheelchair.

With Richard, the nurse he is most comfortable with, Simon braves the intimidating task of learning how to self-catheterize to empty his bladder. He is frightened for days before attempting it. "I don't know what I'm more scared of," he says. "That it will hurt or that it won't hurt." When the time comes, it doesn't hurt, and that is both a worry and a relief.

In the evenings, Simon begins the painful work of trying to play the guitar. Although the acoustic that a friend, John Laird, loans Simon is small and lightweight, at first it is work enough for Simon to practice holding the instrument before he even begins strumming simple patterns. The barre chord requires him to use the full strength of his fingers to fret the strings, and this proves impossible for the left hand. Not using barre chords requires him to rethink how he will play. Simon turns to Guido, who often visits in the evening, to discuss these matters.

"Well, you know there's ways to get around all those barre chords," Guido says. He pulls an acoustic guitar out of a black case and strums a few sample chords. "And there is always the option of playing slide. That would be easier for the left."

"If I start using a slide now, I'll never get the left back," Simon says. He gestures at the guitar Guido holds. "That is a nice guitar," he says. Guitar-envy is plainly evident in his gaze. "Let me see it."

"I picked it up online and got a luthier to adjust the bridge and replace the tuning pegs," Guido says, passing the guitar over.

"Brilliant," Simon says. He strums a G chord, then a C. "Ahh, so sweet."

"It is sweet. The overall sound turned out even better than I expected. I'd love to keep it around, but I got it for you," Guido says. "It's yours."

SUMMER LINGERS THROUGHOUT September. I am grateful for the good weather as I worry that one big autumn windstorm will blow to bits the frail semblance of order we have constructed. I worry that with trees down and roads closed and power out the distance between me and Eli, who is now back at school, will be unbearable. I have moved into the apartment near VGH, the one Guido secured when Simon was still in Step-Down. The fifteen-minute walks between the apartment and GF Strong that bookend my days provide fluid moments of reflection and bridge the busy hours at the rehab center, working so hard to be strong and positive for Simon, and the quiet hours alone in the evening: clutched, knotted, spasmed, anxious hours. These walks, in the lemony light of early morning and the buttery glow of the evening, help me return bit by bit to my own body.

My body is sore. I've lost weight and feel like a ghost, thin and porous, and either teary or numb with lack of sleep. My damaged vertebrae are at the exact same level, T12/L1, as where Simon's spinal cord has been compressed and nearly transected. Always a vulnerable area for me when I am overtired or stressed, my lower right back is in a perpetual state of spasm. I play over and over in my mind the spring nights not so long ago when I was overworked and anxious about my mother's health, when the mild spasm in my back turned into something much worse and I was unable to sleep or move. How I woke Simon to lift me out of bed. The thought that I am not strong enough continues to plague me. How, with my weak back, will I lift Simon? How will I do for Simon what he did so easily for me?

And housing? After we have yearned for a house of our own for so many years, it seems impossible to turn it into a reality. Many people give me counsel to wait: The housing market is at a peak—prices will surely come down over the winter. And it is too soon after a trauma to make huge financial decisions. Who knows what the next few months will bring? The next few years? It is well-intentioned, good advice, but an even louder internal voice tells me that if we—Simon, Eli, and I—are going to survive the next few years, we need a home. We need to not be in perpetual transition. We need a permanent place to begin rebuilding.

I do not find many answers on these walks, only more and more difficult questions, but the act of walking helps. It helps to untangle my thoughts. It helps to reorient me to a world that, for the past month, I have navigated as if half-asleep; everything outside the hospital so bright, so busy, so loud that it hurts. When my questioning thoughts become too clamorous, I tune everything out to concentrate on the idea of sensation or lack of it: What does it feel like not to feel, to be accountable to large expanses of silent skin? I concentrate on the sensations of my skin. The way my cheeks are lit with crisp morning air as Laurel Street stretches itself awake. The rhythm of my gait and the pressure of each reciprocating step, the bottom of my feet inside my shoes, the density and flatness of concrete underfoot versus the energetic cushion of grass and loose dirt. A stubbed toe, a welcome pain.

I RETURN TO the coast to watch Eli's first soccer game of the season and to go on another round of house hunting with Marc while Lorna stays at GF Strong with Simon. It is the weekend and the hospital is quiet. Without his set weekday rehab schedule, Simon is restless. After lunch, he heads down to the gym to do a workout on the pulleys. Lorna, following behind his electric wheelchair, ducks into a washroom on the way, telling him she will join him in the gym in a moment. When she arrives and enters the big gym, Si is the only person there. His back is to her and he is on the far side of the echoey, cavernous space. He leans forward trying to adjust the pulley, a difficult maneuver from his chair. He begins the hard work of a bicep curl, counting his repetitions. He tries so hard to find some improvement, but in that moment his journey toward recovering seems, to Lorna, as huge and lonely as the empty gym. He does another set and then another. He never stops trying. It breaks her heart how he never stops trying.

THE HOUSE ON Cooper Road is Eli's favorite. It is approximately three long country blocks away from our old house and just a few driveways away from the house Simon and I contemplated buying in the spring. From the outside it is a crude, boxy house, a wide rancher with a second-story addition on the left side. It is solid but lacks any gesture toward architectural grace. Inside, the walls are painted in a variety of dark shades—mustard, rust, olive green—all shades I like, but put together, they produce a dark, cavelike feel. Inhabited now by a family of five, it is a busy, cluttered, homey space. The homeowners describe the work they have done on the house, and I see they have significantly increased its value. But there are some design choices that I know Simon will disapprove of. For example, small stones set in black caulking line the tiles of the kitchen countertop, a decorating choice Simon would dismiss with a finishing carpenter's snobbery as belonging to the Blair Witch Project school of design.

But it is big. Big enough for a wheelchair user to not be constantly scraping his knuckles or feeling as if he, with his bulky equipment, is in the way. It is big enough that if Simon needs to fall asleep in his room at five o'clock in the evening, Eli can still have a friend over, upstairs, without disturbing anyone. It has been advertised as wheelchair accessible, but once there, I realize what that means is there is a concrete path around the house and a wheelchair can get inside through the back double doors. To make it fully accessible, we will have to renovate. Our architect-friends Steve and Nikki come to view it.

"With this amount of space," Nikki says, "you have an endless amount of possibility." They immediately offer to help with the design work.

In Simon's absence, his boss, Dave, comes to inspect the quality of construction. "It's well built," he says. "And we'll help you fix it up the way Simon would like."

Lorna hates it. It is dark, and the organization of the space does not make sense to her. She worries that it will require a great deal of renovation to be made accessible, and she wants our transition out of rehab to be seamless and unburdened with extra worry and work. It is difficult to explain that I am choosing this house for what it is not as much as for what it is. It is not perfectly finished: walls need to be torn down and the living-room ceiling raised and drywalled; it needs new paint, lighting fixtures, and appliances. A few skylights and a ramped porch out front. Even though I could not begin to afford this house without financial support from Simon's parents, it otherwise is in harmony with the game plan Simon and I have been working with over the past decade. This is a house still under construction, a house we can move into and, over time, transform. Make our own. Rebuild.

TODAY SIMON AND I are shocked when John, the veterinary surgeon, enters the physio gym upright and walking. He moves slowly and painfully with the aid of a walker and supported by a therapist, but... he is walking. Walking. When I turn back to Simon, I see that he, like me, is crying. He swipes his right hand across his eyes.

"Way to go, John!" Simon says across the gym. Then, to me, more quietly: "I don't think I've ever been happier for someone in my life. Happier, or more jealous." He turns to Sean. "Can you put me on his therapy schedule?"

"Oh, man, I know," Sean sighs. "I wish I could."

Back at Simon's room we learn that Danny is sick. He has been taken to VGH by ambulance and returned to the ICU, a superbug infection swimming in his blood.

"What a day," Simon says. Shortly after seeing John, Simon's physio session had to be cut short because of another condom catheter malfunction, a sudden stain of urine spreading down Simon's gray sweatpants.

"I can handle this. I can get through what has happened to me," Simon says, staring at the empty bed beside him. "But if it had happened to Eli—I don't think I could bear it."

EMILY MAKES THE trip across the country for a brief weekend visit with her wife, Sarah. Simon is anxious. How will he appear to his sister? He explains that sometimes he feels as if he has died and been reborn.

"Reborn. In a wheelchair," he says. "It's hard to explain. I don't know if she'll understand."

"She'll understand." I describe the long nights that Emily and I shared at his bedside in the ICU and Step-Down. "She was there for it all. She'll understand. She'll be ecstatic just to talk to you."

On the airplane, Emily is anxious too. After what has seemed an unbearably long absence she hardly knows what she is walking into. Who will Simon be? And what are the new rules for being with him? Don't look back? Don't look forward? Will it be possible to acknowledge and articulate the profound existential moment that we all collectively have been moving though? What is her new role? Cheerleader? Caregiver? What will she say to him? What will he say to her? She records the moment of arriving at GF Strong in her journal:

SEPTEMBER 14

"I came back."

His eyes were wildly bright when he said that. He was asleep when I first came into his room with Mom and Dad and I was moved to tears by the sight of him: clean-shaven face, hair on his head, mouth dropped open in the classic Paradis sleeping pose— utterly himself. Then his eyes opened and he said, "Hey, Em." I leaned over his bed and we hugged. When I stood up, that's when he said, "I came back," with that kind of frozen-cheeked jester smile people get on acid.

"Thanks," I replied. "I appreciate it." Also smiling.

"So does Kara," he said. "I traveled through time and space to come back so I could be with her."

It took me some time to get used to his new rhythms. A new flatness, hollowness, in his voice—maybe due to the hearing loss? A new intensity, earnestness. Sincere abandon in laughter. Touching honesty about his hopes and wishes, like breaking a silence or entering a conversation with "I'm really hoping my legs come back." He's really hoping his legs come back. Altogether it's a new vulnerability. Of course he's vulnerable—broken, legless, dizzy, wearing diapers, exhausted. But it's something else too. His wit is as quick as ever and he sees beneath what's going on just as penetratingly as before. But there is some meta level of his old thought—a fortress of his thinking about his thoughts about your thoughts about his thoughts—that's quiet right now. In its place, an expanse of compassion.

THE NEXT DAY is game time, Simon's first day trip outside of his daily rehab routine. Arrangements have been made to meet Emily and Sarah at the park not far from GF Strong where Eli's soccer team is playing a match. I have purchased a bike helmet for Simon and preordered a wheelchair-accessible taxi. Simon doesn't appear nervous, just eager to get to the game, so I am surprised to find he is silently crying when I join him and Lorna in front of the nurses' station.

"What is it?" I ask. "Are you worried about the game? Are you worried about seeing everyone?"

"No, no," Simon says, his green eyes veiled with tears. "I just want my legs back and I don't know. I don't know if it's going to happen." He brushes his right hand across his face. "The taxi is waiting." Tears gone, he smiles. "Let's go see a soccer game."

The team manager, Cam, a big-hearted curmudgeon who for years has done the thankless work of ordering uniforms, retrieving soccer balls, and keeping errant boys in line, is the first to meet us, his familiar gruffness a good cover for the tears I see sparkling in the corner of his eyes. He helps us navigate the electric wheelchair over the bumpy sidewalk to a spot on the sideline with a clear view of the soccer pitch. The wind picks up and the cold rain falls with greater determination. Cam rushes off to his truck to exchange my useless little umbrella for a big green canopy to cover Simon.

Emily and Sarah arrive. Sarah, a comedy show producer, was at GF Strong the day before, but I missed her. This is the first time I have seen her since Si's accident, and I too burst into spontaneous and extravagant tears when she hugs me. When I finally look up from her soft, round shoulder, the entire team is lined up, and one by one the boys shake Simon's hand with a solemnity and respect that completely undoes me. We've known these boys since elementary school, and I have never seen them more serious or beautiful. I hang back and lean into Sarah's embrace, not wanting to spoil this moment for Simon with more tears. The boys dedicate the match to Simon and then go out and play the most exceptional game of soccer ever. In a tight, highly competitive league where most games are, as Simon would say, wars of attrition and end in only a goal or two being scored, this game is outstanding. Simon has no problem concentrating or following the complex action. When a speeding ball veers toward him, Cam throws himself in front of Si's chair. Despite the cold and the rain and the rocky terrain, Simon navigates the day without any significant complications.

The final score is 9–0 and doesn't include the two disallowed goals on weak off-side calls. Eli and his co-captain, Nate, lead the team in an inspired game. Eleven times they hit the soccer ball into the net for Simon.

We all return to GF Strong for an impromptu family reunion. Eli changes out of his mud-soaked uniform into cozy sweats, and he and his grandfather go on a foray for takeout; they pick up burgers and fries for all of us except Emily, who has sushi. The evening is cut short when she starts puking.

"Sorry, Si," she says, as she gives him a shaky hug, her face a clammy leaden hue. "Our flight leaves first thing in the morning, so this is good-bye. Sarah and I will bring the kids to visit. Soon. When you are ready for us."

"Love ya, Em."

"I love you too."

SEPTEMBER 22

SIMON WAKES THIS morning so tired he is on the verge of tears.

"I don't know if I can do it. I don't know if I can get up today. I might throw up." His fatigue is excruciating. And it is impossible to get what he most needs—a full night's sleep. "The nurses wake me up every four hours to catheterize," he explains. "Between that, the pain, and Danny's TV, I can't get to sleep. I know he can't sleep either, but who watches *Friends* at midnight!? I fucking hate laugh tracks."

"You could talk to Danny," I say.

"I know, I did. He's a good kid. And the nights are hard. I know when I can't sleep, I feel alone and scared, distant from everything I love. It's a cold wind blowing straight through my heart when you're not here. But ... At least I have the mornings to look forward to." His lopsided smile is resigned. "Except for this morning. I am so fucking tired I don't know how I'll face Sean."

The day does not get any easier. During the physio session Simon's condom catheter slips off, urine staining his sweatpants and wetting the exercise plinth. Sean is relaxed, philosophical ("Don't worry, man. Go back to your room and get changed. We'll pick up where we left off tomorrow"), but Simon is devastated. So upset he can barely speak.

"It's a drag" is all I can think to say, trying to convey in that overused phrase both acknowledgment of the enormity of Simon's emotions (anger, humiliation, exhaustion, despair) and the mundane, practical reality we have no choice but to embrace: equipment failures are likely going to be a part of our life and, as much as possible, we have to deal with them without shame or apology.

Simon doesn't want to talk about it. He wants to be dry and clean. After the long, difficult process of changing out of his sweatpants is over, he asks me to pass him the guitar. He strums the chords to "Hickory Wind," singing along. He has yet to be able to sing it without breaking down on the third verse and once again when he arrives at those lyrics—"It's a hard way to find out that trouble is real / in a faraway city with a faraway feel"—he is overcome with emotion.

"What is it?" I ask.

"Those lines. They just slay me. It's the way I feel," he says. "The loneliness. At night, when you're not here. Like my mind is teetering on the edge of something too big and too lonely to even understand let alone try and explain."

We are interrupted by the appearance of Richard at the door of Simon's room. He takes one look at our sad and disheveled faces and puts his hands on his hips. "Well. You two. I have some good news. Simon, your heparin medication has been canceled and, as soon as your head sore heals up, we can schedule the skull replacement surgery."

We celebrate this news by agreeing Simon should have dinner in bed—a frowned-upon indulgence in rehab—just this once. We eat fruit salad and watch the season's first exhibition game of hockey. The Canucks, fronting a rookie team, win in 4–3 shoot-out.

The following day, Melissa, the OT, requests that I join Simon for his occupational therapy session, and when we arrive at the big gym, she apologizes.

"I hate to do this to you guys," she says, "but we have to go through the pressure sore material."

"Christopher Reeve died of complications from a pressure sore," Simon says as Melissa opens up her reference manual. This piece of information is often traded in conversations at the cafeteria, and the message is clear: pressure sores are kryptonite to a person living with a spinal cord injury.

"They can be really serious," Melissa says. She explains that while a sore may look small at the surface, it may be much larger underneath. She uses the analogy of an iceberg. Pressure starts at the level of the bone, where it forms a base; what you see on the skin is only the tip. She shows us a stage one pressure sore: a dark red blotch of unbroken skin. At stage two, the skin has broken into a small, raw blister. The stage three picture shows a deep crater of mushy, ulcerous tissue.

"That's worse than the pictures on smoke packs," Simon says.

"It gets worse," Melissa says. "I'm sorry, guys."

At stage four, the wound is so deep and open that the underlying bone is visible. Stage four is a horror show of pictures of gaping, puckered holes and blackened dead tissue that parades on until Simon says he has had enough.

"Look, I'm doing a seat lift," he says. Melissa has coached him to shift his weight in the chair throughout the day to relieve pressure.

"Do it every few minutes," she says.

"How about every minute?" Simon presses his hands onto his wheels and lifts his bottom again.

"Even better," Melissa says, then reviews the places on the body that are at the most risk for pressure buildup: the tailbone, hips, sit-bones, and heels. She gives Simon a long-handled mirror designed so that he can lie on his side and view the vulnerable skin on his butt. She instructs him to check twice daily, morning and night.

"Prevention is ultimately much easier than healing a wound," she says. "Sometimes they can get so bad surgery is required."

"You must have seen some bad sores in your travels," I say. Before working at GF Strong, Melissa had traveled with a rehab team to areas of India where people living with spinal cord injuries have little money or access to the types of resources available to Simon.

"Oh yes," she says. Her face momentarily crumples in recollection before her usual bright-eyed smile returns. "Honestly, you don't want to know."

OCTOBER 1

Yesterday I left the Willow St. apartment. Sad, a little, to leave its white saneness. I stayed at the Easter Seal house last night which was loud and not very comfortable—but relatively inexpensive, and I'm not sure where I'll be sleeping tonight. I feel dislocated. Homeless.

In the morning, we had a meeting with Gayle, the sex therapist, who wore a long silk scarf. Typical, I thought, although what do I know? I've never met a sex therapist before. She had an in-your-face sex-positive frankness that was annoying at first but, really, she was charming and informative and very quickly succeeded in wearing down the resistance of my inner prude, the part of me that resented having to sit down with a stranger and talk about our sex life. I now know more about the penis and erections than I could have ever anticipated. Once again we are told that every injury and every body is different. People with higher-up injuries might experience "reflex erections" (the penis becoming erect through the slightest touch, even in a nonsexual context), while people with lower injuries might have mental erections, mild arousal by the thinking of sexy thoughts. Your injury, at L1, is right in the middle, the place where the spinal cord transitions from a unified whole into the "cauda equina," or horse's tail of separate nerves. You might experience either type of erection, Gayle explained, or neither.

So far, you said, pursing your lips, it's neither.

In any case, any type of erection is either created or maintained with the use of a pill, pump, or injection, and it is unlikely, unless there is further mechanical intervention, that it will end in any type of ejaculation. Pills or pumps, Gayle said, should work for you. Some experimentation might be needed. The phrase "use it or lose it" applies here, she said. What's the point of an erection, I wanted to say, if you can't feel it? But I don't. That's something I guess we'll have to figure out on our own. Besides, I can almost 100 percent sure predict she would have said something true, but nonetheless outrageously infuriating, about the brain being the biggest and best sex organ of the body.

Gayle only crossed a line one time—when she told us that some couples report having a BETTER *sex life after a severe spinal cord injury—but I think she knew that was a ridiculous statement to make even before she got to the end of the sentence. You and I just stared at her incredulously. "Hmm," you said, a universe of pained disbelief contained in that single utterance. She gave us both "homework": you, think dirty thoughts; me, take time for myself—get a massage, eat well, read a book.*

Later in the afternoon, Sean took you into the pool for the first time. Your smile, in the water, was saintly, but your mother worried you seemed too childlike.

SINCE THE MIDDLE of September, the newspaper headlines have heralded the decline of the economy, with news of the disintegration of the Lehman Brothers, the largest ever corporate bankruptcy in U.S. history; of Wall Street meltdowns and world market collapses. I, determinedly oblivious to this financial apocalypse, make an offer on the Cooper Street house.

{ 23 }

EYE ON THE PRIZE

OUR FRIENDS STEVE and Nikki give me a key to their city apartment, an attic space I fall in love with. I imagine a starving nineteenth-century artist writing or painting here in romantic, picturesque poverty, complete with porridge and lentil stew, becoming famous only after dying of consumption. It's mine every night of the week except for Tuesdays, when I will share it with Steve and Nikki while they are in Vancouver to take yoga and sculpting classes.

Tonight, though, Simon and I sleep together in the Livingroom for the first time. The Livingroom is a private space on the second floor of GF Strong where couples can book a night together. The purpose, at least in our case, is more practical than romantic, as it gives us a safe place to practice managing Simon's physical care without nursing support. Although its name implies a haven of coziness, I am disappointed to discover that it is no more than an oversized hospital room with a private bathroom and a double bed.

It is a difficult night. Back in his shared room with Danny, who has returned from VGH, thinner and paler but mostly recovered from his infection, Simon and I survive some of the worst moments by telling each other that he is getting better. He *is* getting better, making improvements every day, but what we mean is *all* better. That somehow, by the time he is discharged, he will be well. It is still inconceivable to both of us that Simon won't make a full recovery. But in the Livingroom it is impossible not to confront the reality that certain aspects of Si's injuries are permanent.

"I feel like a beached whale," he says. It is difficult for him to get out of his clothes in the new bed because it is bigger and, unlike his hospital bed, has a grab bar on only one side. The new room is less stuffy than Simon's, and this slight shift in temperature makes his hips ache and spark with neuropathic pain, waking him at regular intervals throughout the long night.

In the morning, we attempt a shower. Simon transfers onto the shower bench, while I mirror the way I have seen Sean support these transfers by steadying the wheelchair and helping with a little extra butt lift. The first transfer goes well, but it is a dry transfer, downhill and to the left, meaning that Simon uses his strong right arm to push off. The transfer back to his chair post-shower is an entirely different matter. Wet, uphill, and to the right, it goes badly. On our first attempt, Simon makes it halfway between the shower bench and chair, but without enough power to complete the transition, we have to retreat. We try again, and this time the chair slides away and Si almost falls. I hold him with my knees bent in a graceless squat, one foot jammed against the slippery wall of the shower, the other against the wheelchair, stuck midway. If I could reach the red emergency buzzer, I would call for help, but if I let go, he'll fall. Somehow, using Hulk strength, I lift Simon into the chair.

"Impressive work, Stan," Simon says. "Thanks."

"No problem," I say, but I am badly shaken by this near-spill. Simon appears relatively untroubled, but I can't shake the sense of imminent disaster that now haunts my day. I recount the gory details to Sean during Si's physio session, but he too is unconcerned.

"If you don't fall a few times—" Sean says.

"—you're not working hard enough," Simon finishes.

"Yeah, well, let's replace the bone flap," I say, "before we embrace falling." My back is sore and I am peevish. I pop Robaxacet and ibuprofen like PEZ candy all afternoon. My head is filled with images I can't shake, of hips cracking and skulls fracturing on hard bathroom tile. I call my mother and tell her I'm afraid I'm starting to think like her.

"Of course, you'll worry. That's natural. But you're doing fine, honey," she says. "You're doing fine."

AS A RESULT of all of Simon's hard work in physio, he eventually transitions out of his electrical chair into a manual one, propelling himself with only the strength of his arms. This is exhausting and challenging work; Simon's shoulders, especially the left, hunch up to his ears with effort. It is almost impossible for him to wheel straight forward; the greater strength of the right arm causes the chair to veer left with every push. In the afternoons Simon returns to the physio gym, angles the left wheel of the chair parallel to one of the bright red lines painted on the gym floor, and practices, over and over again, rolling in a straight line.

After daily care from Richard, the sore on the back of Simon's head finally heals. The heparin is out of his system. The date for his skull replacement surgery is set for the Thursday before Thanksgiving weekend.

OCTOBER 10

We're at VGH. "Unsettling" is much too kind a word to describe the experience of walking down the ICU hallway again. Of waiting hours past the expected time for you to come out of surgery. Of the yellow antiseptic staining the side of your face. Of head bandages and distant eyes and the thick, flat sound of your voice after anesthetic and intubation. Waiting, waiting, waiting.

But Dr. Griesdale's big if has been answered. They have replaced the bone flap. You have survived. This, of course, is the main point, the weight that keeps me grounded. Because it is easy to fall apart here. Because there are always surprises, complications, a sudden turn of events: the discovery you are carrying the superbug MRSA, which increases your risk of serious postsurgical infection; the news that your surgery didn't just involve the bone flap but included an incision under the dura to remove a small clot. Dried blood, Dr. Haw said yesterday. Dried blood accumulated in the left frontal lobe. What that means to your overall health and recovery, exactly, I don't know. Pointless question to ask, though, as I suspect that nobody really knows.

Looks like the Cooper Road house might be ours soon. At the hospital today I am somehow going to have to arrange faxing the paperwork of a counteroffer back to the coast. And secure a room at the Park Inn for Eli and me tonight. It's not even 9:00 a.m. and I am daunted by the prospect of the day.

THE NEURO STEP-DOWN ward is a grim place filled with shrunken old men, one of whom constantly shouts barely intelligible obscenities from his post in the hallway outside his room. There are twisted limbs, useless feet, drool, and ominous pink crinkly balloons tied to bedposts. The walls are covered with posters that chart recovery periods for head injuries and strokes. Small changes might occur in the year following an accident, these charts tell us, but the bulk of healing is done in the first few months.

"God, this chart is so pessimistic," Lorna says. "I can't stand it." We are in the hallway waiting out the nurses' shift change. After the past few busy weeks, it is an uncomfortable déjà vu to be stuck outside Simon's room, anxious and rattled. Marc and Lorna are leaving the next day, and so Lorna is extra rattled. Extra anxious. "What if...," she asks. "What if he is never able to return to a meaningful life?"

I shrug. I am weary and I have no assurances or answers.

"It's so hard..." She stops, overcome. It is desperately difficult for her to leave Simon, and she worries about more than just his health. She worries about his heart. "He's lost so much. I don't think he could stand to lose you." The head injury guidebook contains alarming statistics about the post-injury failure of marriages, and this, in particular, troubles her as she prepares to leave. "Wives come and go, Kara," she says. "But mothers are forever."

For a brief moment I consider being outraged at this statement, but, with consideration, I take no offense. Although we sometimes disagree, over this shared hospital time I have come to depend on Lorna's raw honesty, her courage, and her fierceness. None of us are ready yet to list the positive lessons that have arisen out of Simon's accident, but it is clear that the renewal and deepening of all our family connections is an unqualified good thing. A gift. I know that what Lorna is trying to convey is complicated. She is afraid for Simon, for all he has lost and all he might lose, and acknowledges that there is a possibility—an understandable possibility, in her mind—that I might not accept this new life of ours. And she is telling me, most importantly, that as Simon's mother, she will always be there, willing and able to care for him. She says this not to insult me or question my constancy and commitment, but because these are the preoccupations of a mother, driven to the extremes of worry over her child. Simon is a grown man, but he will always be her child.

The next morning, Thanksgiving Sunday, Simon's face is mightily swollen as if he has smashed into a beehive. The nurses say that it is nothing to worry about, but Marc and Lorna, en route to the airport, worry.

"It feels like a full circle being back here," Marc says. "You guys call anytime, day or night, if you need anything."

"Anytime," Lorna says. "Don't wait."

After Marc and Lorna leave, Simon transfers into his wheelchair, his manual one now, and he and Eli and I go for a roll around the hospital. It turns out to be a morbid tour, revisiting Sassafras and the waiting rooms we haunted in those previous worst of times. Eli is upset and Simon gets a chill. By the time we return to the neuro ward, he is running a mild temperature. Perhaps, we reason, it is the exertion of being up.

Simon returns to bed, and Eli and I make a trip to Capers to score takeout Thanksgiving dinner. We buy too much: pies, soup, cold roasted turkey, glazed brussels sprouts. When we return, Simon's fever is another degree higher, and despite assurances from the nurses, I am destroyed by worry. It is that quick. I hit a wall of despair and fear and exhaustion. Simon is worried too, scared by the threat of infection. I lie and say I know everything is fine. I lie again and say I am going to find a doctor to consult. Instead I escape to the neuro family waiting room and close the door, wishing I could scream, smash, stomp, wail. I sit for ten minutes, holding my head in my hands, and breathe out toxic fumes of panic and frustration before I return to the neuro ward and insist the nurse call a doctor. The doctor looks at Simon's white blood count and says not to worry. I am not much calmer, but the nurse gives Simon a Tylenol, and, as the hospital staff predicted, the fever resolves itself.

It is okay. But I am not okay. I experience the same unadulterated trauma of Simon's first stay at VGH. It sings in my blood. It is not post-traumatic stress. It is not post-anything. It is all still happening.

But—deep breath—there is good news too. My counteroffer on the house is accepted. We now have our very own home to go to.

SIMON ONCE AGAIN rides the transfer van from VGH to GF Strong. The first day back at rehab is hard. I am overwhelmed, preoccupied, and emotionally unavailable as I struggle to figure out all the logistics of making the purchase of the house run smoothly. Simon has five mishaps with his condom catheter during the day; each time he requires a change of bedding and clothing. He calls his father in the evening to tell him about the difficult day, and together they weep on the phone. It is a good release for Simon, and I wish I could join in.

The final condom catheter accident happens in the evening as Simon and I prepare to spend another night together in the Livingroom. I leave to get new, dry sheets for the bed and end up locking myself out of the room, and locking Simon in. Security has to be called to unlock the door so that I can get back in and help Simon out of his wet clothes and bedding. It is a difficult day that ends in a difficult, painfully restless night for both of us. My tears come, late, in the deepest part of the night, when Simon finally falls asleep—hot angry tears that bring little relief.

The next day is better.

"Well, you won't need a Halloween costume next week," Sean says when he sees Simon's impressive skull-spanning incision complete with black, puckered Frankenstein stitches. "You're scary enough as is."

I leave the two of them laughing to visit the bank and sort out our house-buying finances, and while I am gone, Sean— the maniac—takes Simon for an exhilarating ride backward down a flight of stairs in his wheelchair. Sean tells us he was saving that particular physio session for a day when I wasn't around, knowing that I would be a nervous wreck. Next, Simon has music, then occupational therapy and pulleys. Simon has so much energy that he does a second workout with Sean in the afternoon. Then, instead of his usual late-afternoon nap, Simon visits the Paraplegic Association office to chat with one of the peer mentors, Brad.

Although he still needs a great deal of sleep, Simon's overall energy improves dramatically after his skull is replaced, and his social, extroverted nature reasserts itself. There is a shift in his ability to navigate the world: he still wants me close by, but he no longer needs me in quite the same way he did even a week or two before. A new pattern emerges for our weeks: Tuesdays we spend together in the Livingroom, continuing to figure out what is required of us to care for Simon's injured body; Thursdays and Fridays I spend on the coast with Eli, packing and looking for interim housing; Saturdays, Eli is in Vancouver; the rest of the week, while Eli stays with friends, I am at GF Strong during the days and in the artist's garret in the evening. Alone, I luxuriate in the clawfoot bathtub that sits on a platform in the open living-room area before crawling into bed in the small slant-ceilinged attic. There is a novel on the bedside table—Iris Murdoch's hypnotic *The Sea, The Sea*—which I read before I fall asleep. The story is a oceanic swirl of illusion and self-delusion, spirituality and cheap theatrics; I do not retain much of what I read, but while I read, it weaves its magic and for a few brief hours I am insulated from the new fear growing in me.

A DATE HAS been set for Simon's discharge from GF Strong: November 26. It is what he has been working so hard for. It is what we want more than anything. But it is impossible, as it was when he left both the ICU and VGH, to believe that we are ready for what comes next. Simon isn't ready. I'm not ready.

"Our new house needs extensive renos and I haven't found interim housing," I tell Dr. Yao. "Maybe we should push the discharge date back a few weeks? Maybe a month?"

"Hmm," she says. It is clear by the calm look of assessment she turns on me that I am not the first person she has encountered who is terrified to make the transition back home. "Let's leave the date as is. If you haven't found housing before then, we'll discuss options."

OCTOBER 24

I leave you just after breakfast to return to the coast, worried to the point of nausea with the anxiety of leaving. What do I think will happen while I am away? Nothing. Everything. Anything. I don't know. You call just as I reach the Remax office in Sechelt, pen already in hand, ready to sign the stack of paperwork. Your voice is ragged, full of wind and rain. Mom has sent us a care package and in it are pictures from Montreal. It got you thinking about the time you packed up the remnants of the Walker Street apartment solo while Eli and I were in England visiting Mom. It was so lonely, you say. You hate the thought of me packing up Redroofs without you. I'm sorry, you say, I wish I was there to help. Hearing the anguish in your voice I feel my skin flush with tears, the ones that are always so close to the surface, but I fight them back. I don't want to cry in the Remax office, surrounded, as I am, by well-dressed real estate agents. It is tiresome, wearing such a thin skin out into the world. I know it's good, really, necessary and healthy for you to go through the process of feeling this sadness but, God, how I hate to hear you in pain. To hear you in pain and be so far away. This, I guess, was exactly what I was afraid of when I left.

Marooned. This is the word you use. Today, you feel marooned. We talk throughout the day, each of us adrift. Eli and I eat dinner together and then he goes over to Nate's to kick the soccer ball around. It's my first time alone, at home, since the accident. The house feels small and unfamiliar and the air has a distinctly unloved quality to it. I am stunned looking at your guitar, Eli's collection of soccer cleats, the blue bowl on the kitchen table, stunned at how everything looks so clear and sharp but distant.

I am a stranger now in our home. There is so much stuff here, stuff that no longer seems to belong to us. Still, it's true: I can't begin to imagine packing up the last twelve years of our life without you here.

EVERY MORNING I make a list. I focus on executing these chores, checking them off with a smug little tick, with the same single-minded approach with which Simon tackles his daily rehab schedule. Don't look left or right. Don't look back. Just straight ahead.

Keep your eyes on the prize.

This is a phrase I've taken to saying. It is what I tell Lorna when she calls, ravaged by blooming and panicked anxiety, the reasonable by-product of distance and justifiable worry. "How—*how*—are you managing...?" she asks. It is less a question and more an open-ended statement, its implied meaning: there is no possible way you can be managing.

"It's okay," I tell her. "I'm keeping my eyes on the prize."

And later when I express my own panicked anxiety, she kindly echoes it back to me: "Just keep your eyes on the prize, Kara."

It's a ridiculous phrase, the type of motivational jock platitude I usually hate, along the lines of *It's all good.* I am of the particular disposition that believes it is rarely, if ever, *all* good, and when someone tags a conversation in which they are describing a personal issue—an ailing grandparent, a minor conflict at work, car trouble—with a *But it's all good*, I am irritated. How could you say that? Think that? Mean that? But my own ridiculous phrase—*Keep your eyes on the prize*—is a lifeline, powerful as an itemized list, and I regroup, reground, each time I say it.

I do not have a clear vision of what exactly the prize is. But that's okay.

When Simon was in the ICU, I turned my body into a boat, my ribs the ribs of a canoe that floated us downstream to some unknown destination. We're in a sailboat now, one built by many hands, and even though there is no finish line in sight, we are not going to refuse the gift of the wind. This is what I mean every time I say I am keeping my eyes on the prize. Anger, grief, and fear, while a big part of this journey, will not be the anchors we throw overboard, mooring us for God knows how long in some deep, unfathomable place.

WHILE IN REHAB, Simon discovers a previously untapped appetite for reading, and now he is never without a book. We sit and read through the lunch hour, he a copy of *Water for Elephants*, me *The Brain That Changes Itself*, our heads buried in our books to block out the general cacophony of the cafeteria. I am reading Chapter 5, the one that describes an aggressive rehabilitation model for stroke victims, when I have an idea.

"Hey," I say. "Let's spend twenty minutes every day at the end of lunch transcribing a passage out of whatever we are reading using our left hands."

"Why?" Simon is unimpressed. "Lunch is my break time."

"New activities create the potential for plastic change." I hold up Doidge's book and wave it in his face to back myself up. I adlib: "It will help with the guitar playing."

Simon is skeptical but agrees. Over the next few weeks he fills a notebook with oversized and spidery cursive writing, the penmanship of a klutzy seven-year-old. It is an annoying task, but he persists, noting that even in a relatively short amount of time he has gained more strength, control, and connection in his still sleepy left hand.

ONE DAY MIDWEEK, Sean requests that Simon bring his guitar to physio. Simon is suspicious of the request.

"There's no way," he rails as he rolls side to side in his hospital bed, pulling his sweatpants up inch by agonizing inch. "I'm not ready to play in front of people."

"Sean must have a reason," I say. "And it's not like you're going to be performing in front of an audience. Everybody is doing their physio."

"Maybe," Simon says. "We'll see." As he moves through the rest of his morning routine, he decides a few more times that, fuck Sean, he will definitely not take the guitar to the gym, but when it's time to go, he reluctantly agrees to let me carry it down. Just in case.

"Standing-frame day," Sean says and leads Simon over to a large black chair with a hand pump on the side. Simon transfers, on his own now and without a sliding board, onto the chair. Sean adjusts several safety belts and then shows Simon how to pump the handle until he is in a standing position. "You might feel dizzy," Sean says. "Just let me know."

"You're so tall," I say. In the standing frame he is a foot taller than I remember him ever being. He looks like a giant to me.

"I thought you could see how it feels playing while you're upright," Sean says.

"Let me see it," Simon gestures to his new guitar. He plays a few blues progressions, and there are scattered murmurs of approval from the various plinths around the room where people work as Simon has—balancing, rolling over, stretching, doing hand exercises and seated push-ups. The guitar progressions are simple, rudimentary by Simon's previous standards, but his execution is musical and precise.

"Sing something," Sean prods. Simon gives him a dirty look but launches into Bob Dylan's "It Takes a Lot to Laugh, It Takes a Train to Cry."

"Hey," an older man using the stationary hand bike calls out to Simon, "do you know any Elvis?"

And Simon, the reluctant but consummate performer, launches into "That's Alright Mama." Afterward, he plays a few more bars of blues progressions before handing back the guitar and transferring into a seated position.

Later, on our way to dinner, a woman in an electric wheel-chair stops Simon to thank him.

"I don't think I've ever had such a hard day at physio. I was lying on the plinth thinking 'I can't do this, I can't do this' when you started playing," she says. "It helped me get through it, hearing you play."

MARC RETURNS TO Vancouver for Simon's November 26 departure date. Simon and I have been collecting email addresses in an address book and saying our farewells throughout the week so that this final morning isn't a series of emotional good-byes. Sully arrives just after breakfast with his truck, and he and Marc load up Simon's new equipment and the boxes of stuff we have accumulated over the past months while Simon and I do a final, fuss-free, mostly tear-free, parting round, receiving last-minute hugs and advice from the nursing staff and the therapists. I am so thankful to have Marc and Sully there. Without them, I would be swamped by the logistics of this big change; with their support, I am able to be more fully present with Simon. Melissa, the OT, hands Simon a card covered with signatures and warm wishes through the open window of the car, and waves to us as we pull away.

On the coast, on our way to the new Cooper house, we stop for a quick look at one more potential interim housing spot that Joe has found in a small subdivision outside of Sechelt. It's not perfect, but it's the best I've seen yet. I write a check for the first month and damage deposit. We can move in at the beginning of December.

The new Cooper Road house is empty of furniture save for our bed and Eli's, a few boxes of pots, pans, dishes, and towels, and a floor lamp donated by Joe. But the house is warm and well lit when we arrive, and our friend Julie has a feast awaiting us: quinoa salad, sliced beets, cabbage, and steamed greens and—Simon's favorite—a large turkey shepherd's pie. More friends arrive, gradually trickling in. A brief and spontaneous party to celebrate our homecoming ensues. No one stays too long. They toast Simon's return and then leave us, exhausted, to retire to the big empty rooms of our new house.

Deep in the middle of the night. We both lie awake. You say, I'm not the same person you fell in love with. I'm not the same person you married.

I say, yes. Yes, you are.

{ 24 }

Came a Wind

COMING HOME... There is a point after a big trauma when all you want is to get back home. There is also a point when you realize that you will never, ever go home again. That home has now become an evolving state, a rebuilding process, contingent on establishing routines and objectives. The best you can hope for is that "normalcy," whatever shape it now takes, will resume at some point. We'll find our new normal, Kara kept saying.

Despite the fact that I was very fond of my therapists, nurses, and fellow inmates, it was easier to leave GF Strong than I anticipated. The GF Strong staff had done their job well, and I guess I knew it was time to brave life on the outside. It was, however, harder to arrive on the coast than I had imagined. It was intimidating to realize that my physical perspective of this oh so familiar landscape had totally changed.

I wasn't driving yet and so had to ride in the passenger side. I hate the passenger side. Even though we were assigned an accessible spot on the ferry, we quickly realized, as cars filed in beside us, that it would not be accessible for a wheelchair. Plus, even if I could have reached the elevator, the daunting task of doing the sliding board transfer in and out of the car hardly seemed worth the brief time up on deck. So I waited below in the car for the entire ride. As we drove along Highway 101, I saw all the familiar places of my community: the bars and halls I had so frequently played gigs at; the parks I had taken Eli to for soccer practice; the beaches we had swum at and the trails where we had walked Paloma. Ambulating, I had taken access to all these areas for granted. Now I realized returning to many of them would be, at best, a difficult and arduous process; at worst, uncertain or impossible. My left-side weakness made me feel especially vulnerable. One of my goals that I hadn't achieved while at GF Strong was to be able to perform a floor-to-chair transfer. If I fell out of my chair, I would not be able to get back up on my own.

Even with these lingering anxieties, I knew that it was time to get on with the next phase. I had to learn to adapt to my family and social life—whatever that entailed. So, as much as driving up the coast and seeing all of the familiar landmarks was scary, it was also exciting. My sense of smell was gone, but I could imagine the salty air of the ocean and the wet-earth smell of the forest. It was a great and huge feeling to know that I was over the first big hurdle of the health crisis and could now turn the page, so to speak. I felt as if I could reinvent myself accordingly to suit any environment, given the changes I had already navigated in such a short time. As long as there was at least the intention of forward motion, I believed an evolution of self would continue to occur.

Our arrival on the coast announced a whole new beginning. It was terrifying and exhilarating, depending which way the wind blew.

{ 25 }

HAY

SIMON AND I are melting chocolate. Two twenty-pound slabs of milk and dark chocolate dominate the kitchen as they do every year at Christmas. The work of whittling off chunks proves too challenging for Simon's left hand, but he is comfortable parked by the stove, stirring chocolate and, when it is fully softened, tempering it by throwing in an additional handful of meltable flakes, cooling the overall mixture down. He helps me roast almonds, rub the skins off toasted hazelnuts, mix up bowls of dried berries, and cut slivers of crystallized ginger. We mix various combos of nuts and berries into the chocolate and, as I pour the heavenly slurry onto an old cookie tray, Simon uses a spatula to spread it thin. We move it to a cleared spot in the refrigerator and I lick the chocolate-covered spatula clean. Simon laughs.

"This is your happy place, eh, Willy Wonka?" he says.

I laugh too. It's true. Later, when the chocolate has hardened, we will break it up into jagged pieces of bark, fill tissue-lined dollar-store cookie tins, and hand them out to friends. This familiar seasonal routine is both comforting and gloriously indulgent this year in the midst of so much strangeness and difficulty.

We are in interim housing, a fully furnished two-bedroom house on the outskirts of Sechelt, surrounded by the objects of a stranger's life: unfamiliar family photos, vibrating leather recliner chair, a bright red and white art deco kitchen table, quirky can opener and dull vegetable peeler, enormous life-sized TV that takes pride of place in the narrow living room. The rooms are painted in various pastel hues, colors I believe are better suited to Easter eggs than living spaces, and the thick shag rug makes it challenging for Simon to wheel. He inches down the hallway, breaking into a sweat as he moves from bedroom to kitchen. Most difficult, though, is the lack of a truly accessible bathroom. Simon worked with such intent and fearlessness to securely transfer from his wheelchair onto a raised toilet seat, but the bathroom in this house is too narrow for him to continue practicing. Instead he has to use a commode chair, a type of wheelchair that can be rolled over a toilet or into a shower. Our first commode chair is a loaner from the hospital equipment share shed, and it is a disaster. Designed so that a patient has to be pushed, Simon now has to rely on me to get him into the washroom. Wobbly and rusting, the commode chair also has no straps on the footrests to secure Simon's feet, so with the slightest motion his heavy legs thump off, scraping into the rug. I have to crouch down and wheel him by pulling on the footrests so that his feet don't fall off, making the short trip from bed to bathroom excruciating for both of us. When we explain the situation to the WorkSafeBC occupational therapist, she immediately orders us a new and much more functional piece of equipment. Still, the dimensions of the commode chair do not allow Simon to be independent in his morning bathroom routine.

At rehab, figuring out workable bowel and bladder care was the number one concern, not just for Simon but for everyone we met. The topic was discussed daily, openly and frankly, in the hallways and cafeteria at GF Strong. Incontinence offered an odd kind of democracy, and the relief in talking and laughing about it was palpable. *Bowels like routine* was the catchphrase most often repeated as a preface to the discussion of the various rituals that helped an individual's digestive tract navigate paralysis. Simon has avoided most of the drug and suppository regimes that many others require, but he does need digital stimulation to activate his bowels. Because of the dimensions of the bathroom and the commode chair, he cannot reach that far, and so this job falls to me.

"I'm so sorry," he says every morning. "I hate that you have to do this."

"We'll be in our home soon," I tell him as I crouch down beside the toilet. "This is harder for you than it is for me. I'm okay with this." And it's true. I am okay. The ick factor of dealing with bodily issues has long since faded, and our only other option is to have a home care attendant help Simon with these critically important functions. Neither of us want to introduce a stranger into these most private areas of our life. As challenging as the situation is, we are at least able to preserve a modicum of privacy and dignity. Tackle it as a team; stay connected.

It is harder for me to cope with everything else outside of these shared moments. I have been spending a few hours in the afternoon at our old house, moving through what now seems the museum of our previous life, wrapping dishes in tea towels, crying when I stumble over Simon's shoebox of assorted treasures—old cards and letters and concert stubs—or the handmade box, decorated with cut-out butterflies, that Eli made for Simon on his thirtieth birthday, Eli's eight-year-old scrawl across the top: *For Simon's Telecaster Fund.*

Simon sleeps long hours now that he has returned home, but the nights are often fractured by the lurching, jolting pain that wakes him, crying out. After being woken up several times in a row, I often find it hard to return to sleep. Many nights I wait up until four a.m., when I know Emily, three hours ahead in Toronto, will be having her morning coffee and I can call her. It is during one of these early-morning conversations that we both confess to occasionally, guiltily, longing for the clarity, the crystalline quality of the present moment that we existed in when Simon was in ICU.

"Everything is a little muddier now. Weighted," Emily says. "Hectic."

"I know. It's what I imagine it feels like to be a soldier, home from the war but with no talent left for everyday life," I say. "I just want to curl into a ball and sleep for, like, a year."

But there is no time to curl into a ball. In addition to supporting Simon through his daily routines, there is an interim house to run, a new house to plan for, and an old house to pack up. There is a sixteen-year-old desperate to return to his regular family routines. There are doctors to see and physio appointments to make. There is the ever-present pressure that the window will close on Simon's recovery time. And then there is Christmas. We all take a break for Christmas.

Snow falls throughout December, unusually thick and consistent for our rainy coastline. Snow banks accumulate at the side of the road, making it difficult for Simon to go for a wheel around the neighborhood. But the world outside the bedroom window is dream-lit, edged in silver, flaked with white, and the close, dense blanket of clouds is a comforting inky blue: a perfect Christmas snow globe. Our new next-door neighbor, Guy, hangs lights on our house and a wreath on our door, and one Saturday afternoon Eli goes out with Cam, the manager of the soccer team, and returns with a beautiful, bushy fir and a tree stand. I retrieve a box of our accumulated decorations. Up and bejeweled with lights and baubles, the Christmas tree is at least one familiar fixture in the house. Simon and I make ginger-bread cookies, chocolate bark, and cranberry sauce, enjoying the sense of imposed snow-day confinement. Eli, however, is rarely at the house. He is ecstatic about the more-than-usual snowfall and the unexpected income opportunity it presents in the form of shoveling. It is Eli who suggests purchasing a Wii console for Christmas.

"For Simon," he says and blushes. "It'll get him moving a bit in the chair."

"Yeah, you probably wouldn't play it at all," I tease.

"Well," he says, still red-faced, "it is something he and I can do together."

The Wii is a hit. Christmas morning, Simon and Eli play several rounds of electronic golf while I stuff the turkey and coat it in a sage and pancetta butter. Mom arrives midday with her two dogs and three oversized Christmas stockings, one for each of us. She wanted to have Christmas in the new house on Cooper, but I refused. Even without the snowbound roads, it would have been too much work to transition between houses.

"I stopped at Cooper on the way through," she says as we all sit down for dinner. "It's going to be great. I really wish we could have had Christmas there. It's your first house. It's a big deal."

"Next year, I promise." We snap open the red crackers. Simon reads his joke out loud while Mom and I don our green paper crowns. "At least we're not buying cold takeout to eat in Simon's hospital room like Eli and I did at Thanksgiving."

"Yeah," Eli grumbles. "No leftovers."

"And no more chemo either," I say. Mom's hair is now a fine silver bristle covering her whole head.

"To surviving," she says and we all raise our glasses, clink-ing before we dig in. "Your gravy is better than mine," she adds, her mouth full. It is less a compliment and more an accusation. Nobody messes with my mother and her gravy. She sighs, smiles, and raises a turkey-laden fork: "Well, honey, I guess it's time I pass on that gravy torch."

JANUARY IS A difficult month. Simon, being so dependent, is acutely aware of every pulse, every shade of my shifting moods, and often I am ashamed as he witnesses how frequently my will falters. He knows the exact moment when I want to give up, go back to bed, and pull the covers over my head.

"Oh, Stan. I'm so sorry all this work lands on you."

"Don't apologize," I say. "It's okay." But it is not okay. I am frequently overwhelmed by the daily chores and the work needed to care for his body. Even my lists can't save me. I am daunted by the idea of rebuilding our lives, which seems a futile pursuit, impossible and illusory, now that we live with the sure knowledge that everything can change or be destroyed in the span of a few seconds. I might as well be trying to construct a three-bedroom family home out of pieces of driftwood gathered on the beach.

I worry constantly that we are not doing enough rehab or are wasting the "critical healing time." I book Simon for weekly physiotherapy sessions at the hospital, though they depress him. I find them depressing too. After a month of nesting and watching DVDs, it is hard to face the reality all over again of how sore and depleted Simon's body is and how compromised his left-side strength is. And the physio room at the hospital is depressing all on its own. Small and understaffed, it is full of elderly, mostly overweight and radically underdressed people, missing limbs, suffering from strokes, or fighting for some postsurgical mobility. Signs are posted all over the hospital advising people to avoid the building if at all possible because of an outbreak of Norwalk flu—Simon's nightmare. Jennifer, the physiotherapist, is long-limbed and athletic, and she and Simon get along well, but often she manages two or more patients at the same time, and there are long stretches where Simon lies on a plinth halfheartedly lifting weights, trying to avoid the invisible flu germs. When she does work with him, she is indignant about his lack of progress.

"No floor-to-chair transfers yet?" she exclaims. "When I worked at GF you wouldn't have left rehab without being able to do one. And what's with the anti-tippers on the back of your chair?"

The anti-tippers are two rods with small caster wheels that act as a backup balancing system. New wheelchair users often remove these wheels, as without them the chair can more easily traverse uneven surfaces. But because Simon's balance is so challenged and we are afraid of another fall, not only has he left his anti-tippers on, but he still wears his seatbelt at all times.

"I know Jennifer is just trying to motivate me, but I can't get with the whole macho para thing," Simon says after his first session. "I want to be healthy, but I'm no Rick Hansen. I'm keeping my safety belt and anti-tippers on, and I'm not going to turn into some weight-pumping gym dude now that I'm in a chair. I wasn't that guy before my accident."

"I hear you," I say.

Despite Si's lack of enthusiasm, we continue to go. It's good exercise at least to get out of the house and practice the difficult transfers in and out of the car.

SIMON'S FEET BECOME massively swollen, and every night I massage them when he gets into bed, trying to encourage some blood flow back up his legs. Then one Sunday morning at the end of January, he discovers a sore while getting dressed.

"What the fuck," he hollers. From the kitchen I hear the tone of someone's heart dropping to the bottom of their gut and I come running. "What the fuck is that?" he says, pointing at the image of his right foot in the mirror adjacent to our bed.

There is a two- by three-centimeter oval of purple necrotic tissue on his heel. A deep tissue wound. We call the nurse at WorkSafeBC who came by only two days ago to inspect the swelling in Simon's feet and legs. She instructs us to take pictures of the sore and outline its shape with black marker so that we can chart whether it is getting bigger or smaller. A battery of visits and consultations with doctors, nurses, occupational therapists, and physiotherapists follows, each of whom has slightly different suggestions, though the basic principle remains the same: allow no pressure on the heel at all.

Simon and I make a twenty-four-hour plan. We promise each other we will be diligent and resolute until it heals. I lie and say it will be fine. Simon agrees with my lie and says, yes, it will be fine. But the truth is that skin this thin is the ultimate betrayal, and the idea of a pressure sore flattens us both. It robs me of all my remaining energy, and I spend days running to the bathroom every few hours to crouch or cry or vomit. I am horrified that I didn't notice the sore before it got so bad. Fear and doubt about my ability to cope, unlike anything I have ever experienced, worms its way into my heart. The kind nurse at WorkSafeBC tells me over and over not to blame myself, that a pressure sore can happen terrifyingly fast—in less than four hours. Despite her good intentions, this fact brings no relief.

SIMON'S FOOT HEALS, slowly but, thankfully, without any sign of infection. Our days become a little busier and more structured. We begin to develop some consistent routines and rituals, and they fill our days like a full-time job. Simon's daily bathroom, shower, and dressing routine takes the entire morning to accomplish, and then, just before lunch, I help him with a series of stretches and range-of-motion and balancing exercises. The afternoons vary according to the day of the week. On Wednesday afternoons, Simon visits with the psychologist, Bruce. Based on my reading of *The Brain That Changes Itself,* I encourage Simon to try out new skills, so on Mondays he takes an hour-long lesson with a classical guitar teacher, Paul, and on Fridays I drive him to the Sechelt Art Centre for a figure-drawing class. Tuesdays and Thursdays, I return to a three-o'clock Pilates class in an attempt to decompress my cranky lower back while Joe Stanton visits with Simon. Saturday afternoons we watch Eli play soccer, and Sunday is our rest day: we snack on mandarins and popcorn and watch the boxed set of The Wire DVDs that Simon's cousin Lia sent us.

I also buy the Posit Science brain-training program featured in Doidge's book, and it becomes a daily afternoon activity. There are two discrete programs—one that targets the auditory cortex; the other, the visual cortex—and Simon spends an hour on each every day. The auditory cortex program begins by differentiating pitches and various phonemes—a *ba* and a *da,* say—and progresses to various memory games. The visual cortex games are more colorful and varied and work at increasing the speed at which visual detail is processed. Simon is occasionally frustrated by these games, especially differentiating *ba*'s from *da*'s, and he curses the software in a way that is out of character for him, but he perseveres. We eat dinner by five-thirty because by six, Simon is bone-tired and ready to get back into bed. This combination of brief but intense physical activity (getting through his morning routine and exercises), mental activity (the classical guitar, drawing lessons, and brain games), and extra sleep is extremely productive, and almost daily I notice improvements in his overall focus and clarity of thought.

Joe's visits are always comforting. Joe doesn't like to talk a lot; he just comes to play. "Hey, Simon," he says when he arrives. In my mind, the way Joe says "Hey" is always spelled with an a: Hay, Simon. "Hay, Joe," Simon returns, mimicking Joe's ageless cowboy drawl. Joe's wisdom and grace lie in knowing that when words fail, as they so often do, there is always music, and he coaxes Simon into a lyrical and rhythmic conversation. With an intuitive sense of when to push and when to pull back, Joe provides a framework for Simon to explore musical thoughts and ideas. At first Simon can play along only for a few minutes, but gradually their playing time lengthens, and soon they start practicing specific songs. Often, returning home from Pilates, I hide around the corner of the kitchen in the carpeted hallway so I don't interrupt. I call Emily, holding up the phone, so that she and I can listen to them sing together.

The Precious Littles delayed having the official CD release party after Simon's accident, but a date has finally been set for the end of February, and one day Joe asks if Simon wants to play a song with the band.

"I'd like to do it," Simon says after Joe leaves. "But . . . I don't know if I can."

"Why not?" Eli says as he sets the dinner table. "Why couldn't you do it?"

"I don't know," Simon says. "I can't play like I used to. I'm scared, I guess."

"That's the wrong way to look at it, man," Eli says. "You've already stared down the scariest things possible. So. Nothing scares you now. Of course you can do it."

"I don't know," Simon says. But he continues to practice, and three weeks later he makes his first public appearance at the sold-out CD release party at a local art gallery. Guido is filling in Simon's spot on electric guitar for the gig, but Simon joins the band for the opening number, "99 Days," a song Simon describes as having tumbleweed blowing through the hole in its heart.

I hope my baby's still waiting for me on the Outside
I dream of her arms around me
But these bars surround me and
I've cried.
99 days
Never been so long
Every man pays
pays for what he's done
It's no way to live, no way to die
It's no way to live, no way to die

The entire band plays on a raised platform except for Simon, who chooses to park in front of the stage. That way, when he is done playing, he can simply wheel away without the disruption of having to be lifted off. He keeps his playing simple, echoing the riff he played on the baritone guitar for the recording. His left hand is tight, and fumbly when he makes quick changes. He is nervous, yes, but he is doing what only four months ago would have been an impossible feat: he is playing live with his band. I lean in, hard, to the joy of this moment, along with everyone else in the room.

{26}

A CELEBRATION

THE SNOW MELTS. Spring is on its way, but work on the new house has yet to begin. The process of approving the renovation plans and hiring a contractor is much longer and more involved than I had anticipated. Stuck in interim housing, we chip away at each day, but sometimes, especially early in the morning, my impatience overwhelms me and I want my life back—my work, my things, my pre-accident daily routines—in all of its busy, messy, mobile, chaotic permutations. This small, narrow house in a West Sechelt subdivision, with its shag carpet and mauve walls, is the perfect setting for an updated version of Sartre's *No Exit*. I would title it *Marooned,* borrowing Simon's word. I am ashamed of these thoughts because, as much as I chafe at the slowness and smallness of our life, I am, unlike Simon, still free in my body. Despite the far more radical transformation his life has undergone, he rarely complains. The other day while Joe was over and I was out buying groceries, Simon's condom catheter came off. Instead of their regular music session, Simon had to enlist Joe in helping him clean a puddle of urine off the kitchen floor. When self-pity burbles in my heart, I remind myself, that that was a hard day.

But my thoughts are full of big holes, unbridgeable gaps. How to convey the constant daily contradiction of emotions, battling? How to convey the constant courage needed to face the day? How courage is needed moment by moment but how, instead, it comes and goes, vaporous, dissipating with the grind of repetitive chores.

This morning Simon wakes up angry. Angry at his unpredictable bladder. Angry at his unpredictable bowels. Angry that his spinal cord injury never progressed from an ASIA A to an ASIA B or C. There is nothing I can say to ease his anger and ... why should I? He deserves a day to rage.

It is all over the afternoon news that the actress Natasha Richardson has died of a traumatic brain injury. She fell on a beginner's ski slope outside of Montreal. She refused ambulance assistance at the time of the fall but a few hours later became unstable. She was taken to a local hospital, then Montreal, then New York, but it was too late. Pressure on the brain stem caused brain death, and eventually they took her off life support. I am driving when I first hear the story and I have to pull the car over. I sit on the side of the road and weep for her and her family and I weep for us, for what we went through, and how close we came to a similar ending.

Over dinner, our sad, angry day behind us, we discuss what kind of stove would be most accessible. I have my heart set on a fancy range with a fully accessible side-opening oven door. Perhaps it is a little true that happiness can be bought: after our difficult day, it thrills me to think about buying our first ever stove for our first ever house.

SPRING ARRIVES EARLY and with it a series of unqualified good things. The pots of strawberry plants on the back patio grow green and bushy, and hummingbirds frequent the feeder outside the kitchen, flashes of iridescent rust and peacock blue and kiwi green reflecting in the window. I get my first work contract since the accident, doing structural editing on a children's book. Eli arrives back home from a soccer trip to Spain and France world-weary and years more grown up. Simon's beloved Canucks secure a spot in the playoffs. Work begins, slowly, on the renovations at Cooper. And, at the beginning of April, I take Simon out to the medical offices at UBC for a neuropsychological (cognitive skills) test requested by WorkSafeBC.

Simon is nervous but immediately establishes a rapport with the psychologist, a woman named Lisa, who is about our age. It is a full day of testing, and I am not allowed to remain in the room once the assessment begins. I wander the UBC campus, one of my favorite places, and return at scheduled breaks with tea, sandwiches, and chocolate chip cookies. While Simon completes a written portion of the test, Lisa takes me into an adjacent room. She asks what changes or deficits I've noticed. I hate this type of question and feel as if, simply by answering, I am betraying Simon. But I answer as honestly as possible. I don't see any big gaps in Simon's cognition or personality, but... he is tired all the time, I tell her, and often in pain. He is slow in his thinking and speaking, and his balance is poor. The Posit Science games do cause a level of irritability and frustration that is unusual for him. I have worked hard to keep his recovery period stress-free but I suspect that if he had to deal with any basic level of responsibility, like cooking regularly or paying bills, both his stress and frustration levels would skyrocket. When he is extra-tired, or in a busy environment, there are moments when his logic, memory, or tactful nature might falter.

"For example?" she asks.

"At a soccer game," I say. "When my son's team played in the division finals a few weeks ago, it was a very intense and exciting game. As the action heated up, Simon, on the sidelines, shouted at the opposing team players. Nothing outrageous—the type of stuff he regularly yells at the TV when the Canucks are playing—but not really tactful at a kid's game. And it's not like him to do that." I hate saying this out loud—this is the type of observation I avoid writing in my journal—but it is also a relief to let it out. "It's an intermittent issue. Not a consistent problem at all. Most often he is his usual tactful, sensitive self. But when it does happen, I don't know how to broach it. I don't want to make him self-conscious, and I don't want to mommy or nag at him. Do I let it go? Or do I say something?"

"I've always found head-on and honest works best," Lisa says and smiles. "And it's important to know that what is happening is not a problem per se with any particular mental faculty—memory, logic, or personality—it's a problem of the brain becoming overloaded when he is tired or overwhelmed. I would imagine this will get better with time." She smiles again. "You know, it's a good brain he has. His test scores are exceedingly high. Given the severity of his injuries and the reports I have from GF Strong, I wasn't expecting to see what I'm seeing in there."

At the end of the day, Simon is exhausted but duly rewarded by Lisa's summary of his test scores.

"Your baseline score was high. Very high. I'd place you somewhere in the 98th percentile of the population, in a range deemed very superior intellect. Now, there are a few areas where you score much lower—motor speed, visual-spatial problem solving, and expressive language—and this would indicate areas that were damaged by the head injury. But I'd like to point out three things. First, although there is a significant discrepancy between your highest and lowest scores, your lowest scores are all within an average functioning range. So a head injury for you looks like where many of us are at to start with. Second, in some of these low-score areas, you did eventually come up with the right answer but just not in the allotted amount of time, so I couldn't score you points. This suggests one of two things to me. Either these areas will improve with time or you will have to compensate by taking more time to accomplish certain tasks. Third, you are emotionally and psychologically strong. For someone with test scores as high as yours, I would expect to see a certain level of neurotic tension in dealing with a severe head injury. Panic at not remembering a word or experiencing difficulty problem solving. Your sense of humor and general approach seem to provide you with the necessary tools to navigate your recovery with amazing grace."

Simon smiles sideways at me, eyebrows raised. Whoa.

"Is there anything I can do to help those injured areas continue to heal?" he asks the doctor.

"You guys just keep doing what you are doing," she says. "Other than that, I would give you the advice I'd give anyone. Keep trying new things. Learn a foreign language. Travel. I know you have a lot of complicating health issues, but when you're ready, you are going to be able to do anything you want."

"Except snowboard," Simon says.

"Except snowboard," the doctor agrees.

"I told you," I say on the way out to the car. "New things!"

"Shush," Simon says. "I get to be the unchallenged smartypants for today. Oh man, I can't wait to tell my mom."

I lie awake in bed that night beside a softly snoring Simon and let go of all my lingering doubts about his cognitive functioning. I am not particularly surprised by the test results, as I've always known that Simon's intellect was fierce: strong, flexible, and versatile. I am more surprised at how relieved and validated I feel. Since the moment Simon woke up, I have felt him all there—challenged, yes, and struggling to connect all the parts of his thinking, but all there. All Simon. Still, after so many different doctors sat me down and told me that no matter

what potential healing took place, the severity of Simon's injuries meant his brain would never be as good as it once was, I
questioned my perceptions. Were there gaps in Simon's thinking that I was willfully oblivious to? Did my love and lifetime
of shared experiences conspire to fill in these gaps? Did my
brain actively work to re-create the post-accident Simon in
the image of the pre-accident Simon? And if so, was that, ultimately, detrimental and unfair to him? For the first time, I feel
as if can accept what the doctors have repeatedly told me: his
brain might never totally regain its previous speed or stamina
or balance. And while the cognitive test results don't make the
loss of speed, stamina, and balance irrelevant or unimportant, they do underscore my conviction that Simon has plenty
of strong, healthy gray matter capable of re-creating its way in
the world. He doesn't need me to fill in any gaps.

ONE AFTERNOON WHILE Simon has his classical guitar lesson, my mother and I go out for lunch. It is a year since she
finished her last chemo treatment. Her hair has returned to its
familiar curly tufts, and her latest blood test indicated she is
cancer-free. She is ecstatic about Simon's test results and my
work contract.

"I've been so worried," she says as she hugs me. "But you
guys are going to be fine." We make plans to meet the following
week at the same time to go to a nursery together. She wants to
buy a flowering cherry tree to plant in her front yard, but when
the day rolls around, I ask Simon to call her and cancel: I am
sick with a head cold and have lost my voice. Three days later, a
Tuesday, I call, my voice still weak and scratchy. A high school
friend has recently published a novel, and Mom and I spend a
happy half-hour trying to match the fictional characters with
their real-life counterparts. She calls back a few hours later
while I am in the bath and leaves a message: "Listen, honey. I

just heard a news story on CBC. I think maybe you have swine
flu. Call me."

The next day, Wednesday, Eli calls to tell her he passed
his license test and is the proud owner of an N sticker, but
she doesn't pick up. We call again the following day from the
ferry. The three of us have spent the afternoon at GF Strong,
where Simon, along with four other recipients, was presented
the 2009 Gert Vorsteher award, an award to celebrate remarkable determination in pursuing rehabilitation goals. I regret not
having invited Mom—she would have loved the ceremony—
but once again she doesn't pick up. Friday I work a long day
and by Saturday I have completed work on my editing contract.
I am searching for the phone receiver in the papers piled on the
kitchen table when it rings. I assume it's Mom calling, as I was
about to call her to make plans for Mother's Day tomorrow. But
it's not Mom. It's Joe Stanton's partner, Sue, with news. A Powell River RCMP officer contacted the Halfmoon Bay General
Store trying to get an accurate house address for me. The store
owner called Sue, who now gives me the number for the RCMP
officer. Even before I dial, I know with every atom in my body
what the news will be.

MAY 10, 2009. MOTHER'S DAY

*Last night I received news that Mom died. She died at home in her
bed, most likely in her sleep, which is a blessing.*

*Wednesday, they think. I unwind the week in my mind. Was
that the night that Eli and I both inexplicably woke up, meeting
in the kitchen at three, each of us drinking a tall glass of water? I
remember thinking that it was the type of sudden unexplainable
wakefulness that I associate with the night I found out my father
died. I remember thinking something bad has happened or something bad is about to happen. I almost said it out loud, to Eli, but
stopped myself, thinking that's the last thing he needs. Thinking I*

don't need to catastrophize everything. Thinking, God, I'm turning into my mother.

She's been on my mind more than usual this week. And now she's dead. Death. Gone. She is so much an integral part of me: her tall bulk, her big love, her holding too tight, her generosity, her need. I never wanted her to be alone.

This is the incomprehensible thing: her heart stopped. Red pepper pumping organ, weakened by chemo and radiation and smoking and a lifetime of stress. Si says I loved her and she drove me crazy. And now there is a big empty house full of her stuff. I am scared of that house without her in it. I am scared it is going to smell like death. She was there, dead, alone, for at least two days. There are going to be so many pictures and books and clothes. Boxes of crackers and bunches of celery. How could someone living alone buy so many big bunches of celery? How do I plan a funeral?

I RUPTURE. I spill outside myself. Eli and Simon watch in horror as I puddle, barely able to stand for the waves of grief that roll through me. I ask Emily to come and, despite her very busy schedule, she arrives the next day to spend the weekend. She stays with Simon and Eli, filling our freezer with veggie lasagnas. My brother flies in from Singapore, and, together, he and I return to Mom's big empty house.

THE DAY EMILY leaves, Simon, Eli, and I travel up to Powell River for Mom's service. Simon is desperately anxious about having to navigate a new, and possibly inaccessible, environment for the day. Eli is quiet, attentive, and extra helpful, but reserved. My brother and I have adopted the popular word *celebration*—today is a *celebration* of Mom's life—but I am afraid, as we travel up to Powell River, that I have no more celebration left in me. We gather at the Laughing Oyster, a local restaurant owned by Mom's friends. Despite the short notice, many people arrive: clients, co-workers, community members, my mother's best friend, Sylvia, who has flown across the country. My mother would have been delighted to have so many of her friends gathered together in one room, and, to my surprise, I find a spark of celebration. To honor my mother's great appetite, her extravagant love of food and drink, my brother has ordered an open wine bar and a buffet of exquisite hors d'oeuvres. Sylvia speaks of her more than three-decades-long friendship with Mom. I read Charlie Smith's poem "The Meaning of Birds." After, we all go to the beach where she walked her dogs every day. Eli, Simon, and I, along with Mom's friends, build tiny paper boats and place lit candles in them. The ocean is calm, a textured expanse of luminous gray with sky-blue tufts of cumulous clouds scrunched and heaped along the horizon, backlit by the heavenly glow of the setting sun. Gulls and ravens wheel in the silvery light while Mom's dogs, Molly and Finnegan, race along the pebbled beach, chasing the birds that dare to land. Simon's wheelchair can't navigate a route down to the water's edge, so he parks at the top of the trail and sings "I Shall Be Released" as our convoy of white lights float out to sea.

The next day, as Simon and I travel back up to Powell River, he begins to shake, his torso shuddering so violently against the car seat that I think at first he is having a seizure. No, he tells me, he is cold. His skin turns ashen and he can't stop shivering. His teeth chatter and his hands clench the car door. We are on the ferry, and I wonder if I should call for an ambulance to meet us on the other side. But, given the distance from the ferry terminal to town, I decide against it. I can get Simon to the hospital faster if I take him myself.

We wait for hours in the emergency room until finally a doctor explains that Simon has a serious kidney infection. He wants to admit Simon to the hospital. This doesn't seem possible to me. Eli will be back at school tomorrow. Simon and I cannot remain in Powell River while Eli is on the coast, not after Mom has just died: it is an equation that simply doesn't work. I sign a form stating I am taking Simon home against doctor's orders, promising the anxious ER doctor that Simon and I will return to the hospital in Sechelt.

By the time we reach the coast, it is after nine at night and our local ER is filled with casualties from the annual longboard contest, fit young men and women with large patches of flayed skin or arm bones jutting unnaturally from their bodies. Simon is given a dose of an IV antibiotic, and we are told to return to the acute care ward once a day for the next ten days. We return to the strange rooms of interim housing just before midnight. Simon is dizzy with exhaustion and I am numb: too tired to be sad or too sad to be tired. After Simon is safely in bed, I write a final entry in my journal. But it is the last one I write for a very long time.

Mom. There are a few things I wanted to tell you.

1. *I cut my hair the week after you and I went out for lunch. It's short. Okay, but pouffier than I wanted. It reminds me a little of your hair when we lived in Orangeville but, unlike you, I don't fight the curls.*

2. *Eli got his license. He was so proud of himself, and you were the first person he wanted to call.*

3. *I don't have swine flu but Simon is really sick. I am scared and worried every minute of the day. Most of the time I have no idea what to do.*

4. *Simon was embarrassed after receiving the Gert Vorsteher award because in his acceptance speech he thanked the GF Strong nursing*

staff of the second floor when, of course, he spent his entire stay on the third. More than anyone, I know you would have loved the ceremony. It was a day to honor those who are truly inspiring and courageous and you would have been very proud of Simon.

5. *I'm sorry we didn't go to the nursery and pick our spring bulbs and seeds together.*

6. *They have finally started renos on the Cooper house. Oak floors and fir doors! We are very excited, but moving in still feels very far away.*

7. *I was thinking about you all week.*

8. *I was worried the cancer might come back. I was worried if it did I wouldn't be able to be there for you. I'm happy that you weren't sick for a long time and I hope you weren't scared when it was time to go.*

9. *Your garden is beautiful this spring.*

10. *I love you.*

{ 27 }

WECANDOITWECANDOIT-
WECANDOIT

———————————

THE MOMENT WHEN Eli arrived in the world was the most densely real moment of my life. Simon's, too. So real that for a brief time Eli outshone everything beyond the reach of his radiant newborn energy. The world outside faltered in its frequency, all the newspapers and lawsuits and stock options and grocery stores less real and substantial than Eli's tiny perfect fingertips, his wide-open eyes and hungry mouth. The moment I walked into the glass room in the ICU and saw Simon, hooked from every part of his body to various machines and missing a third of his skull, was an equivalently dense moment, a moment so real that everything else that followed became unreal in comparison. Life continued to play out its shadowy dance, but now we were only images flickering in and out of focus on the wall of Plato's cave, a reflection of something that was once more real.

So even something as simple as brushing my teeth becomes an act I perform for the person staring back at me in the mirror, a sensation that is even more acute when I watch Simon.

It is in his reflection, as he shaves or flosses, that I see the effects of the stroke on the left side of his face. What I routinely don't notice—the lack of mobility and the droop of his lip, the extra-wide opening of his eye, the slight sag to his cheek—is shockingly clear when I look at his reflection. More real? I want to ask him. Or more unreal? But I don't. It's not a question sensible people ask out loud, and the real answer is a non sequitur. The real answer is this: something shifted in all our spirits after Mom died. After months and months of living in a heightened state of awareness, we are all beyond exhausted by the extremes of our happiness and our sorrow, swells of emotion that are less like riding a roller coaster or surfing colossal waves and more like being the patient in a violent game of Operation, our hearts routinely plucked from our chests only to be dropped back in again.

Simon finishes his round of IV antibiotics. There are no further extreme complications from the infection, but neither does he recover completely. He is plagued by an uninterrupted series of infections in his urinary tract and begins taking an almost daily dose of antibiotics. He and I travel to Vancouver, where he takes a two-week-long driver retraining course, learning to drive a car with only hand controls, and re-earns his license. Eli finishes grade 11 and gets a job washing dishes.

The renos proceed slowly but are far enough along that in June, a week after Simon's thirty-ninth birthday, we move into the house. Dave's team has done a transformative job. Along with the agreed-upon renos to make the house accessible, they have added skylights, wood floors, and fir trim: the house is big and bright, open and beautiful. Many of the tradespeople who worked on it donated costly supplies and labor. The accessibility renovations immediately and profoundly change our day-to-day existence, as Simon is now more capable of being independent in his showering, dressing, and bathroom routines. It is a good, good thing, a great gift, this house. Eli, Simon, and I know this, but it is more difficult to *feel* the excitement and gratitude. The move derails the simple but nourishing schedule we established in interim housing, and although we are relieved to be settled in our permanent residence, this new house, no matter how bright, beautiful, and accessible, is not home yet. The empty rooms are filled with the packed-up contents from both my mother's house and our Hobbit house, and we all move over and around the unopened boxes like quiet ghosts. Eli is less help than he would normally be. When he is not working, he prefers to escape to the beach with his friends Jesse and Nate. I don't begrudge him the fluid world of the ocean, and besides, I have no energy to argue or insist he do anything. Simon, tired of apologizing for not being more help, escapes to his own world, reading novels and playing cashless games of online poker. The work I do opening each box, sorting through the remnants of my mother's and our previous lives, is slow-going and treacherous; a molasses feeling of fatigue settles into my body. I know that depression would be an accurate way of naming our collective exhaustion, irritation, and numbness, but I see no point in doing so. For the moment, exhaustion, irritation, and numbness seem a reasonable response to the past eleven months. There is nothing to do right now but to get up each day and keep moving through it as best as we can.

Emily, Sarah, and their kids, along with Marc and Lorna, arrive on the coast the third week of July. It is the first time Oscar, who is nine, and Alice, who is six, have seen Simon since his accident, and their approach of loud, rambunctious, open-hearted curiosity is a gift to our quiet house. Lorna is amazed by the house's transformation and immediately retracts all of her reservations.

"It's beautiful," she says. "You two have done an amazing job."

On the twenty-second, the first-year anniversary of Simon's accident, Emily and Sarah cook up a feast, and we invite Joe Stanton and his partner, Sue, over. A second bottle of wine is opened after dinner and Marc suggests bringing out guitars. Joe retrieves his from the car but Simon declines to play. It is seven o'clock on a sunny summer day but he is sapped of energy. So am I. We leave our guests on the back patio and retreat to our bedroom.

"There's no anniversary to mark," Simon says as I help him into bed. "What happened a year ago—it hasn't stopped. It's all still happening for me and you, isn't it? It's never going to stop."

OSCAR AND I walk to the corner store to buy small brown bags of penny candy. Born and raised in Toronto, he is delighted by the ducks and deer and raccoons that populate our semirural streets. He points to the thick brambles dotted with pinkish-white flowers that line the side of the roads.

"Are those raspberry bushes?" he asks.

"Blackberry," I say. "In a few weeks they will be covered in big, juicy blackberries, and when the weather gets hot, the road will start to smell like cotton candy. Or hot jam."

"Hot jam?" he says, eyes shining. "Oh. My. God. I can't wait to smell that."

I nod. "Me too."

Ten days after they arrive, the family leaves. During their visit, a great deal of work has been accomplished. Lorna hung pictures, adjusted furniture, and filled vases with artful arrangements of flowers. Marc completed a series of odd jobs—touching up paint, organizing the garden shed, flushing out the eaves of windfall and construction debris. Emily helped establish a triage order to the unpacked boxes, opening what needed to be opened and storing what could wait. Sarah filled our freezer with brisket shepherd's pie. The kids laughed and tussled and had water fights with Eli, Paloma chasing after them with an oversized stick in her mouth. It is impossible to pinpoint the exact moment, but sometime during their visit our big, beautiful house has become a home.

The day after everyone leaves, Dave delivers the piano that Simon arranged to buy for my birthday last year, and every evening for the following week Simon teaches me chords to familiar songs: "Hickory Wind" or Bob Dylan's "Like a Rolling Stone." During the day, I take Simon to physio sessions at the hospital, where he practices standing upright, braces on his legs and balancing on parallel bars. The workout is challenging but exciting. Theoretically, this work could lead to Simon's being able to stand upright and "walk" short periods using crutches. However, this eventuality is unlikely given the great probability of falling, and therefore the risk of another brain injury. But it is good, exciting, to see Simon working out right on the edge of his ability, pushing his limits. It is a great week, one that feels full of the promise of new beginnings, but it all ends when, on Sunday night, I check Simon's feet at bedtime. While the family was visiting, Simon had an appointment with a podiatrist to address the problem of ingrown toenails. The podiatrist removed a wedge-shaped piece of the outer edge of Simon's big toenail on each foot. The left toe healed nicely, but the right toe is slow to close up and now looks infected, the skin darkening to a blackish purple.

"It looks like the pressure sore on my heel, doesn't it?" Simon asks.

I agree, and even though it is Sunday night I leave a message at the office of the very amiable and approachable podiatrist. "Simon's right toe isn't healing. I think we need to see you as soon as possible."

But in the morning the blackness on the toe has continued to grow; the whole round tip is mushy with pus. And it continues to grow at an alarming rate, doubling in size by the time we reach the ER, and smelling sickly sweet, like rotted fruit.

"It's necrotic," the surgeon tells Simon. "We'll have to admit you. I'll try and save the toe."

Simon is in the hospital for a week, and everything is so terrible it is almost funny. Almost. First there is the diagnosis: a superbug MRSA infection in the toe, which will require twice-daily doses of the end-of-the-line intravenous antibiotic vancomycin for six weeks to two months. Second there is the accessibility, or lack of it, in our local hospital. Not a single washroom in the rooms on the ward is wheelchair accessible, meaning that Simon must perform his daily constitutional in a commode chair in the middle of the room. As in interim housing, this setup does not allow Simon to be fully independent: he once again needs assistance to get through his morning routine. Third, the mid-August weather, as if in accordance with our mood, turns unseasonably bleak; an early autumn chill descends, and the clouds lower and grow heavy, as if at any moment they might fall from the sky and cover us in a suffocating blanket. Our new home is drafty, cold, and lonely when I return in the evening without Simon.

We push to have Simon discharged early, but there are logistical problems. He needs the antibiotics twice a day, precisely twelve hours apart, a process that takes two hours each time. It is not reasonable either for Simon to remain hospitalized or for us to visit the acute care unit twice a day for the next two months. So the surgeon puts a PICC (peripherally inserted central catheter) line into Simon's upper right arm that leads straight to his heart, and I am taught how to hook up the IV antibiotics. We return home with a bag full of vancomycin, an IV stand, a box full of medical supplies, and, at least for the moment, Simon's toe intact. Simon is giddy with relief to be home, but I am unnerved by the prospect of living with an invisible superbug, especially when Simon has an open portal from his arm straight to his heart. I spend my days scrubbing: hands, sinks, door handles, Simon's shoes.

THIS SUMMER MARKS the last year Eli will be eligible to play in the annual Chilliwack soccer tournament. As we approach the end of August, Simon reminds me of the promise I made last year.

"We have to go," he says.

I agree, but the idea of traveling to Chilliwack is daunting. Because Simon's health is so precarious, we arrange for Eli to travel with his best friend, Jesse, and, despite my reservations, I go ahead and book a room at a Cozy Court, the only accessible room I can find. I release some of my neurotic exhaust by calling the hotel several times to clarify how the rooms are set up.

"There are grab bars by the toilet, right?" I ask. "On both sides?"

This time last year, getting to Chilliwack was an impossibility. This year, we can do it. When I find myself overwhelmed with doubt about the wisdom of making the trip, I repeat this phrase over and over, like the Little Engine That Could: *wecandoitwecandoitwecandoit*. But right from the start, nothing goes well. The day before we are to leave is my fortieth birthday, and that morning Simon wakes me up with shocking news.

"There's blood in the bed," he says, pushing himself into an upright position. "Oh. Oh no. The PICC line has come out. I must have snagged it in my sleep."

As I get up to investigate the bright red stain spreading across our white sheets, Simon starts to breath rapidly. "I feel dizzy," he says, his skin pasty. "Light-headed. Like I might pass out."

I rush for the phone to dial 911, recalling the nurse's words when the PICC line was inserted: if Simon felt light-headed or short of breath, we should get medical help immediately. By the time the ambulance arrives, Simon's breath has returned to normal and I have realized that the blood stain, while dramatic-looking, does not constitute a large amount of lost blood. Embarrassed, I apologize to the ambulance attendants who arrive.

"If you are questioning whether you should call an ambulance or not, then it's a good idea to call," the female attendant says, smiling. "I'm only glad it's not more serious." They give Simon a thorough check and okay him to proceed with his morning routine at home. He and I drive to the hospital just before lunch to meet with the surgeon.

"We're lucky," the surgeon says pointing to the small cap on the line that is threaded into the heart. "It's still intact. If it had come loose when the line came out, we'd be calling a vascular specialist right now." He would like to insert another line, this time in Simon's left arm, but if he does this, we can't go to Chilliwack, as Simon will have to return to the hospital twenty-four hours later for a checkup. We compromise by making an appointment for Monday to have the PICC line inserted, and a nurse puts in a temporary butterfly line into Simon's forearm. Once again we leave the hospital with our bags of IV vancomycin and a box of supplies, as well as a stamped letter attesting to the fact that Simon's IV line is for therapeutic—not recreational—drug use, just in case we end up in another ER.

We leave for Chilliwack early on Saturday morning and have just enough time to check in to our hotel before Eli's first game. "Oh," the desk clerk says when she sees Simon's wheelchair, "I just filled the accessible room."

"What?" I am aghast. "I called you guys, like, four times to confirm we had an accessible room and to ask you specific questions about the setup."

"Well, you could just use a regular room, right? The wheelchair would just have to enter from the patio doors."

"Can the wheelchair get in the bathroom? Are there grab bars on the walls?" It is unusual for me to yell—throughout our relationship Simon has usually taken on that job—but I am straight-up shouting. "Look. We have just traveled more than five hours and we have nowhere else to go. We need that room. We need it tonight."

Yelling, as I'm sure most yellers would attest, proves a very effective method of getting what you want. The desk clerk places a call to the newly checked-in inhabitants of the accessible room. Elderly and slow-moving but not disabled, they readily agree to switch. As I unpack our car, my back, tight and achy since I called the ambulance the day before, suddenly starts to spasm so fiercely I can barely straighten up enough to drive to the soccer pitch. I hold onto the back of Simon's wheelchair and try to stretch it out as the ref's whistle announces the start of the game and the red versus white jerseys thunder past, feet flying and torsos twisting as the kids, agile as gazelles, leap into the air for headers. It is a new team for Eli this year, and the players are not yet accustomed to working with one another. The afternoon matches go poorly. They lose 5-1 in their first game and are shut out 3-0 in the next. When the games are over, it is too late for Simon and me to join the team for the annual meal at Earls. I give Eli money for dinner, and he leaves with Jesse and his father, while Simon and I return to the hotel to start his medication.

The IV stand didn't fit in the car, so I rig up the bag of vancomycin on a coat hanger and hook it over the shower curtain rod, but when I connect the IV nothing works. Simon has blown the vein transferring in and out of the car. I get directions to the hospital from the front desk and we leave, my back complaining loudly as I dismantle Simon's wheelchair so that it fits in the trunk. The ER waiting room is packed, mostly with people in various states of inebriation, and a posted sign states that the expected wait time is four hours. Shortly after we arrive, a very large, very drunk man stumbles into the waiting room holding a hot dog. He lies down across the three remaining seats, places his hot dog beside him on the tiled floor, and falls asleep.

"I hope he's not planning on eating that when he wakes up," Simon says. He is genuinely worried. "Someone has got to stop him from eating that hot dog. It's not sanitary."

An hour later, a drunken scuffle over a missing cigarette lighter breaks out beside us in the waiting room. Simon turns to me, eyes glassy, and says, "This might be the worst night of my life."

"That's saying a lot," I say, "considering the year you've just had." I have to do something. When a nurse opens the locked door to call the next patient in, I barge into the treatment rooms.

"He doesn't need to see a doctor," I say. "We just need a nurse to set up the butterfly line. He needs to get on his meds; we're already two hours past when he should have started. I have a letter." I brandish the signed note from our hospital and, thankfully, this works: a nurse leads us into a treatment room and finds a new vein.

It is after eleven o'clock when we return to the hotel room. As I once more attempt to set up the IV, I realize with dismay that the butterfly line attachment from the Chilliwack ER is completely different from the one used by our hospital. Try as I might, I cannot figure out how it works. The vancomycin has to slowly drip into Simon's veins, and I worry that if I do something wrong, the liquid will rush into his system and cause a toxic reaction. I call the hospital, and the nurse we saw tries to walk me through the process. But my nerves have gotten the best of me, and I am so terrified that I will do something wrong that I can do nothing at all.

"Can you come back in?" the nurse asks. "I'll walk you through it."

And so I leave Simon in the hotel room and return once more to the hospital, where the kind nurse gives me a tutorial in butterfly IVs.

"Got it?" he asks.

I nod. It is a totally straightforward process, and I am embarrassed that I couldn't figure it out back at the hotel room.

"Don't worry," the nurse says. "It's not so easy when you've never done it before."

It is close to one when I limp back into the hotel room. I get the drip started, finally, as Simon, weepy with exhaustion, droops in his chair. To pass the two hours needed to safely drip the vancomycin into his bloodstream, we discuss the feasibility of a reality show that visits the ERs of towns across Canada at twelve o'clock on a Saturday night.

"You could learn a lot about a town that way," Simon says.

"An extreme reality show?" I say. "Or extreme unreality?"

"Lord," Simon says, driven to uncharacteristic religiosity, "please let me make it through this night."

A lot of Robaxacet, a brief sleep, and two more hours of the vancomycin drip later, we return to Eli's soccer games. It is another frustrating day on the pitch. Eli's team ties one game but loses the next two. Simon and I leave Chilliwack just before the final quarter of the last game. Emily calls as we arrive home, and Simon and I regale her with all the gory details of our trip. Now that it is over, it seems both epic and comic in its awfulness. Home, finally, our exhaustion dissipates into something closer to triumph than tragedy.

"By the way, happy birthday, Stan," Emily says.

"If there ever was a weekend that could embody how the downside of forty should feel," I say, "this was it: bad back, blood in the bed, an unholy trinity of trips to the ER."

"Oh, guys," she says. "I'm so sorry it was that horrendous."

"It sucked from top to bottom," I say. "But we did it. I'm so glad we did it."

SIMON'S HEALTH ISSUES dominate our life through the fall. Twice daily we repeat the time-consuming vancomycin and toe wound care routine. Almost daily there are medical appointments: bloodwork and bone scans, visits with our GP, the surgeon, or a wound care nurse. Almost every other week there is a visit to the ER: even on the vancomycin, Simon continues to get infections in his urinary tract, and every hint of a fever sends us scuttling to the hospital. Simon jokes that I am putting in my practicum hours as an URN, an unregistered nurse, but I do not feel very competent. I desperately mistrust my ability to read the severity of what is happening medically. I knew, for example, that Simon's right toe was not healing as quickly as the left, but still, I thought it was all under control. Conversely, I panicked when the PICC line came out and when I was setting up the IV in the hotel room. The presence of a superbug completely undermines my ability to accurately gauge what a situation requires. The round of vancomycin ends the second week of October, but by November the infection in Simon's toe has returned. More vancomycin, the surgeon tells us. This time for twice as long.

"You're kidding me, right?" Simon says. "There's got to be another option."

The surgeon takes in our hostile faces, both of us reacting as if MRSA is something he personally cooked up for the sole purpose of torturing us.

"I'm on your side, guys," he says. "You know that, right?"

We leave his office and stop at the grocery store to pick up dinner supplies. Simon stops wheeling and rolls to a halt by the coffee grinder.

"If this is it. If this is my life—infection, vanco, more infection, more vanco," he says, looking up at me, "I have to wonder why? Why would the doctors work so hard to save me, if this is it?"

"Oh, Beau," I say as we slowly start back down the aisle.

"I'm sorry, Stan," he says. "I know that's not what you need to hear."

"It's okay," I say. "I get it." I remind him of Bill, a guy we met briefly at GF Strong. He had suffered a complete thoracic injury in a car accident, and nine months after his initial surgery, he was returned to the operating room because the hardware inserted to support his spinal column had become infected. He told us he had been okay up until that point, but when he learned of the infection, all the potential depression that had been lying in wait descended on him, and he barely had the energy to eat or get out of bed.

"I don't think I'm depressed," Simon says. "It can't really be depression when you are reacting to things that genuinely suck ass. I'm not depressed, I'm low. I'm blah. You know—*blah,* depression's distant and more ironic cousin."

It is costly to us, the time Simon is on vancomycin. We are more withdrawn and have little community interaction. The amount of time spent on health-related issues precludes any other activities. Even attending our one established weekly outing—Eli's soccer games—is not always possible, and weeks go by when the only people we see are paid medical staff. In October, I start teaching again at our local Capilano University campus. It is only one two-hour class for six weeks, but, although I look forward to each class, it becomes increasingly stressful to leave the house.

Ever since his time at GF Strong, Simon has experienced a certain level of neuropathic pain. It is, for the most part, not achy and consistent but rather a vivid, searing jolt of pain.

"Like someone has hooked battery cables to my hips," Simon says, "and they're shocking me." Now, three months into the vancomycin regime, his pain level suddenly skyrockets. It comes in big, continuous waves and leaves him howling, gripping a table, unable to move his wheelchair. It is terrible when he is sitting but worse when he lies down, and it goes on for hours and hours.

"It's like living with a wolf with its leg caught in a trap," Eli says one morning after a particularly bad night. I call the department co-ordinator at Capilano and tell her I have to cancel my proposed classes for the winter and spring. It is too soon for me to return to work.

There are more tests and X-rays, but no explanation for this increase in pain is found. Our GP prescribes a variety of medications, some of which significantly affect Simon's level of consciousness, none of which are totally effective against the pain. So he is dozy, sometimes to the point of hallucinating, but still in pain. One night he wakes me up, insisting I go turn off his computer.

"Don't worry," I tell him. "It's off."

"No, no," he says. "The email hits, they're jolting my hips."

I get up and assure him his computer's off. I put on the kettle, resigned to another sleepless night. I make tea and think how the days and weeks of dealing with the vancomycin regime are transforming us all. Eli, normally a healthy and active person, has been beset with sore throats, fevers, and stomach upsets. He is anxious and short-tempered, as am I. I worry constantly that Simon is one fever away from returning to the ICU. It is a life-and-death worry, not unlike what I felt when he was actually in the ICU, the kind of worry that, at some level, we are all familiar with but that is not meant to be sustained through the daily activities of shopping, cooking, cleaning, working. And Simon, fearless Simon, is afraid of the pain. When it starts up, he physically cowers. It breaks my heart. It is amazing, I think, how he could wake up from the massive trauma of a central nervous system injury and still be himself. All of us, in that moment of trauma, and despite the shock, were all still recognizable: Kara, Eli, Simon. It is the erosive forces of chronic pain, worry, and infection that threaten to turn us into something we were never meant to become.

{ 28 }

A HUNDRED MILES

JANUARY 2010

IN THE NEW year Joe Stanton and the bass player, Gerry Millar, come to the house for weekly rehearsals. Joe has been asked to play an original song at the Olympic Torch Relay concert in February, and, if Simon feels ready, he will join Joe and Gerry to form a Precious Littles trio for the gig. The PICC line has made it uncomfortable for Simon to play guitar, and he has practiced very little over the past six months. He is self-conscious about his abilities on both his electric and acoustic guitars and so chooses an instrument he rarely played before his accident, his metal resonator guitar. In between vancomycin drips and hospital trips, he begins to work on his parts.

Things have not changed dramatically from the previous month. Simon's new cocktail of pain medications seems to work marginally better, though his pain still routinely reaches epic, horrifying levels. The vancomycin regime is still as time-consuming, but it has become easier now that it is more routine. Eli still has a perpetual ache in his gut, and I am still worried.

But the days are a little longer, a little lighter, and the Torch Relay Concert and the Olympics are something to look forward to. Even though Simon is critical of his playing, I remind him that this time last year, he played a single song at the Precious Littles' CD release. Then, he could barely keep up, strumming the simple baritone riff to "99 Days." Now he will be able to play through and solo on "Make Me Proud." Although the past year felt like a series of baby steps forward followed by several huge slides backward, his playing has progressed.

A LARGE STAGE has been erected behind the courthouse and a wheelchair lift has been installed so that Simon can access it. On the day of the concert, the crowd that swells out front is huge, one of the largest Simon has ever played in front of. More than a thousand people, someone tells me. Almost three thousand, someone else says. Whatever the actual number, it is a big crowd.

"Kara," Joe says when we meet up with him backstage. "I'm nervous." It's true; he looks pale and a little breathless. Gerry, too.

"Hey, guys," Gerry says. He lays the large case of his upright bass down and pushes his owlish glasses a little farther up the bridge of his nose. Simon, however, is beaming.

"Hey, Simon," Joe says, in his slow drawl. "Aren't you even a little jittery?"

"Yeah, I am," Simon says, "but mostly I'm just happy to be here and be a part of the show."

I am more nervous than Simon, I think, when he rolls out onto the stage. There is a bit of jockeying around as he tries to navigate the electronic gear and coils of cables, but Joe and Gerry help him get set up with his resonator guitar. Joe picks up his acoustic and perches on his stool, and Gerry stands beside his upright bass. The emcee, a CBC television announcer, introduces the band, noting that this gig marks Simon's return to public performing since a work accident a year and a half ago.

Backstage, I sing along when they reach the chorus: *I'm going to sing like the rain on the roof. I will find my truth and I will make me proud.* The applause when they are done is thunderous. Simon rolls off, beaming still.

"Make me proud," Gerry says, reaching out and giving me a bear hug. "You bet."

"You did good, Si," Joe says.

"Thanks, Captain Little," Simon says. "It felt good. It felt like coming home."

MUSIC MAKING, OR musicking, is one of the most complex cognitive challenges the human mind can undertake. Neurobiologists have demonstrated that the act of making music activates the executive functions of the prefrontal cortex, the emotional centers of the limbic system, and the motor systems in our basal ganglia and cerebellum, linking the most highly developed, specifically human functions of the brain with the most ancient and animal-like. There is likely no other single activity that activates as many complex neural networks while at the same time tickling the brain structures involved in motivation, reward, and emotion. This is why we can perform the difficult task of remembering the melody and lyrics of a long-ago song—say, a song to remember the French names of body parts or one played at a grade eight dance or sung by our mother—with relative ease.

Research compiled by Gottfried Schlaug and his team at Harvard has shown that the complex combination of sensory, cognitive, and motor functions needed to make music is a strong stimulator for brain plasticity in both the developing and the adult brain. They have shown that not only is the front part of the corpus callosum (the area of the brain that connects the left and right hemisphere) larger in lifelong practicing musicians, there are also microstructural changes in the cerebellum. This ongoing research suggests not only that music is a powerful therapeutic tool for people recovering from brain injuries but that, in general, musicians appear to be less susceptible to age-related cognitive decline, presumably as a direct result of their daily musical activities. Schlaug writes that "music might also provide an alternative entry into a 'broken' brain system to remediate impaired neural processes or neural connections by engaging and linking up brain centers that would otherwise not be engaged or linked with each other."

This is what I think about when I read that passage: Marc, my beautiful father-in-law, sitting beside his son for days singing the repertoire of family songs by Bob Dylan, the Band, and the Beatles. Simon is strapped into an upright position, his head droopy and body listing to one side, his eyes unfocused and sleepy, his left hand clenched in a tight claw. Marc launches into "Hickory Wind," and on the third verse Simon suddenly lifts his head and looks toward his father. He mouths the shape of words: *It's a hard way to find out that trouble is real.* The next day he begins writing words; two days later he is speaking. I know it is egregiously unscientific for me to make such a bold and unprovable claim, but I believe this to be true: music opened wide the closed doors in Simon's mind.

THINGS DON'T MAGICALLY fall into place after the Torch Relay gig, but it does mark a change in our life. The vancomycin antibiotic regime finally ends, and Joe and Simon sit down to review a list of potential gig dates for the next four months. It's not a lot—a gig or two a month—but it's both paid work and something to work toward. Simon is resistant at first but finally forgoes practicing on the resonator guitar in favor of his preferred instrument, a tobacco sunburst Fender Telecaster with a rosewood neck. I return to my writing, finally finishing the rewrite of my graduate thesis project. Eli works two jobs over the summer, saving for the fall, when he will start at the University of Victoria in the Health Sciences Department. In our backyard we build raised garden beds and pathways for the wheelchair so that Simon can plant tentacled potatoes and rows of sweet pea seeds. As green things start to grow, it is possible to look back and marvel at the formidable moments we survived and the progress we have made. Continue to make. And it is possible, if only for a moment, to feel transformed into something blooming and hopeful once again.

The promise of a new normal was a promise we had made in good faith, and now we are able to start discovering it.

"Cool," Eli says when I tell him I am planning to write a non-fiction account of Simon's accident. "But—" he adds, a deep wrinkle furrowing his brow. "How does it end?"

SEPTEMBER 2010

OVER THE LABOR Day weekend Guido, Simon, and I drive Eli to the bunny-infested campus of the University of Victoria and help unload his laptop, duvet, and mini-fridge into his dorm room. He is excited and nervous, anxious, once we have helped him unpack and hugged him good-bye, for us to hit the road so that he can embark on this new chapter in his life.

"I thought things would be so different when we reached this point," Simon says on the drive home.

I agree. As young parents we had fantasized about how young we would still be when Eli left home. How, in our early forties, we would do some of the things we hadn't been able to do in our twenties. Travel. Eat cereal for dinner. Stay up all night. Sleep in all morning.

"Instead," Simon says, "we're living like retired ninety-year-olds."

It's true. Back at home, the house feels hollow without Eli's presence. I miss the swarm of teenage boys that regularly moved like locusts through my kitchen in the early afternoon, devouring milk, bread, and bananas on their way up to Eli's room. Without that youthful energy, our life settles into a very prescribed routine. To help Simon manage his bowels, we eat a carefully planned diet at specific times of the day. We unplug the phone for long stretches of time to help Simon get the sleep he needs. We don't indulge in even the occasional, casual glass of wine. We monitor our sugar intake and watch too much TV. Staying up together until ten o'clock is a big deal. Retired ninety-year-olds is about right.

JANUARY 2011

"COME LISTEN TO this," Simon calls from his office. He's playing the promo mix CD that came in his Christmas-stocking Mojo magazine. "I can't believe I've never heard this cut."

"In a minute."

Simon turns up the music and Bob Dylan's voice, full of grit and longing, fills the kitchen while I finish cutting onions for a pasta sauce, wipe the knife and cutting board, and wash my hands. For a Dylan fan like Simon, an unknown song is indeed a rarity. And this is vintage Bob: young, acoustic, mournful. A coyote whistling at the moon.

In his office Simon leans forward in his wheelchair and I think he is in the grip of neuropathic pain, but when he leans back I see it is not a spasm that contorts his face; it is tears. He has cried only a few times during his recovery, and never like this. His face is as wet as a child's in the pouring rain.

"What?" I kneel beside the chair. "Are you okay?"

"It's the song. We were so young when we left home too," he says through the tears. "You know?"

"Play it again." I sit on the floor. He restarts the song as I lean into his wheels. "*I can't go home this a-way, this a-way, lord, this a-way,*" Dylan sings. "*I can't go home this a-way.*"

"Do you want to go back to your parents' place? See the rest of the family?" I ask. It is a difficult, nerve-wracking task, reintroducing his new self to the world. Everything that was once easy is now hard. The idea of the air travel back to Quebec seems an impossibility.

I know the answer, but I ask the question anyway. "Are you worried about the prospect of going back there?"

"Yes, but that's not it." He shrugs off my literal interpretation. "It's here…" He cues the CD and sings along. "*You will hear that whistle blow a hundred miles—*" His voice cuts short with a choked sob. He rubs his eyes with the heel of his hands then blows his nose before continuing: "It's the way he sings 'a hundred miles'—I don't know. It makes me realize there's no going home anymore. That there is no home to go to. Never again. Not really."

I know what he means. I want to say, It's okay, babe. It will be all right. But I don't. As much as I want them to stop, I know these tears are good, cleansing. It just kills me to hear him cry.

THIS IS OUR new reality: things aren't easy. The everyday difficulties of life are often made more difficult by the demands of Simon's body. Infection is a constantly recurring issue. Pain is a daily violence—a violence and a puzzle. Simon and I look for patterns, try to isolate causes, but to no avail. The tools and resources we both have previously used to solve problems, often with a good deal of success, fail us utterly here. Playing music makes the pain more tolerable, but holding the guitar for too long brings it on. The pain acts up when Simon has spent a long day in the chair, but it gets even worse when he stretches out in bed. Stress aggravates the pain but, it seems, so does a deep, relaxed sleep. The very worst is late at night, except, of course, for the days when it starts first thing in the morning and is bad all day. The difficult truth is that the pain is as random and quixotic as bolts of lightning. Simon has a preternaturally high pain threshold, but some days the pain defeats him, defeats us both.

The medical system's solution is to write a prescription. Chemical intervention is not what Simon wants, and it often muddies his mind, makes him sleepy, and clogs up his already temperamental digestive system. But compared with the screaming, electric pain that shoots through his lower back, hips, and right leg, it is the lesser of evils. Alternative therapies and breathing exercises bring little to no relief. The question Simon is often asked—sometimes well-meaning, sometimes condescending and judgmental—only compounds our frustration. "Hydromorphone is addictive, you know. And it plugs up the system," an emergency-room nurse informs Simon in a tone of moral outrage. "Why are you *still* taking it?"

"For pain," Simon says. He is unnaturally subdued, feverish, and exhausted after waiting almost three hours to see a doctor to get a prescription for a urinary tract infection. Later he tells me he wanted to say, "For fun. I'm taking it for fun. It's all part of my carefree rock 'n' roll lifestyle, lady." But he didn't. He was worried that he would get red-flagged as a potential substance abuser.

The other question we are often asked is whether the cause of the pain might be emotional. This question is most often asked by concerned friends and acquaintances and arises out of the popular belief that certain pain in the body can be related to specific emotional issues—something like the attempts of the phrenologists to attribute certain character traits to a particular shape of a skull. Low backache, we are told, means you have financial worries or sexual woes and are generally taking too much on. As much as I believe in the interconnectedness of our emotional, physical, and spiritual selves, this literal translation of that connection, especially in the context of Simon's injury, is absurd, offensive. The question itself, like the one the snooty nurse asked about hydromorphone, belittles the stark reality of a pain that is so real and raw that it is its own separate entity, a snaking, many-clawed beast that moves though Simon's body at will.

The confusion about the type of refractory (meaning "not yielding, or not yielding readily, to treatment") chronic pain Simon experiences is understandable, though, because it is, quite simply, confusing. Confounding. Like most people, Simon and I have previously experienced only acute pain, pain that is localized and that triggers the nervous system to alert us to the possibility of injury or disease. Acute pain has an immediate and specific message: Pay attention. Now! Something is wrong.

The message of chronic pain common to people with spinal cord injuries is not so easy to decipher. Known as either spinal cord, central, or neuropathic pain syndrome, this pain is a neurological condition that affects the central nervous system, which is composed of the brain, the brain stem, and the spinal cord, pain that is often felt for those that are paralyzed, with cruel irony, at or below the level of injury. Within a healthy spinal cord, pain signals are directed to various regions of the brain to be processed, but when the spinal cord itself is damaged, communication within the central nervous system can become abnormal, the intricate interplay of the naturally occurring chemicals being disrupted by trauma or disease. In the damaged spinal cord, pain signals—transmitted up the spinal cord to the thalamus, the double walnut-shaped region of the brain that processes pain—increase. Spinal cord, or central, pain is, like acute pain, an alarm, but in this instance the alarm doesn't switch off; instead, it communicates pain to the thalamus in unrelenting bursts, experienced by the sufferer as electrical jolts or a shooting, burning, clawing, splintering sensation. The more this alarm sounds, the stronger and louder it gets. The biochemical changes that occur in the central nervous system and allow for this magnification of the alarm signal can persist years after the original injury has healed.

Simon's strategy is not to dwell on it but, during the pain-free times, to get down to the business of playing music. It is a good strategy and it serves him well. I, however, spend a great deal of time thinking about pain. At its worst, when it comes in wave after unending wave, Simon's pain recalls the extreme state that accompanies the final throes of a fatal cancer, but for Simon this is pain he has to learn not to die with but to live with. It seems too much for him to endure. It is a state of being that gives the lie to positive thinking. How do you maintain the fantasy that everything will turn out okay when there is a daily abundance of evidence that everything is not okay? As much as I try to understand the mechanics of chronic pain, neither of us can override the urgent message of our lifelong experience of acute pain: When pain is that severe, we kick into high alert. Something—somewhere—must be wrong.

In an attempt to change my thinking, I visualize Simon's spinal column as an erratically sparking, severed electrical cord. I imagine wrapping it securely with a roll of duct tape. I remind myself, and Simon, that his body has done a magnificent job of healing profound injuries. It has healed deep wounds. Yet there is this small section of his spinal cord, no more than an inch or two, that has not been able to finish its healing. The story I tell myself is this: The erratic sparks of this ruptured electrical cord, the pain that it causes, is a manifestation of his body's attempt still to heal the trauma of the central nervous system; it is a manifestation of the body's biological imperative toward healing and wholeness. The body keeps trying. Pain means Simon's body is still trying.

OUR ANTIDOTE TO a really tough day is bluegrass. When either of us is feeling low, blah, Simon pulls out the guitar. While I steam spinach or grill chicken, architecting an early evening dinner, Simon serenades me from the kitchen table. For many years, out of shyness and insecurity, I muzzled my sing-a-long voice to an atonal whisper, worried that if I raised the volume, I would surely honk or bleat out a variety of bad notes. But now Simon coaxes me to sing out loud, stopping the song occasionally to help me find the right note. Together, at the top of our lungs, we sing Southern spirituals, bluesy laments, and songs of love and heartache until we feel whole again. I have no illusions: I am an enthusiastic singer but not a good one. Simon tells me the more enthusiastic I become, the more my notes skate and slide in and out of key, but those brief moments when he and I know we are in perfect harmony—well, those moments are like real heaven to me.

THE FIRST FULL gig Simon plays is at the Garden Bay Pub. Ron, the owner, who came to visit Simon often at rehab, has upgraded his establishment by building wheelchair-accessible washrooms, an act of inclusion that means the world to Simon and in a small way mitigates his nervousness about playing a full three sets. At Simon's request, Joe and I have not publicized to either our friends or the wider public the fact that Simon will be playing. I glance around the room as the band sets up and recognize a few regular faces, but not many. The bar is only half full. Simon is nervous and chatters away like a magpie as they proceed with a sound check, Joe occasionally nodding his head and inserting a *yep* or *Hmhmmm*.

They open with Dylan's "Watching the River Flow" and follow up with the title track from the Precious Littles' CD, "Sometimes You Win." Simon's chatter stops the moment the set starts, as he concentrates on some of the standard riffs he used to play on Joe's originals. He is focused on watching his fingers move across the fret board and does not look out at the audience. It is a marked contrast to his pre-accident demeanor; Simon was a natural performer who exuded a kind of playful largesse, an ease and humor that charmed audiences. Now he struggles to keep time with the band. But he does. And then he sings a song, Taj Mahal's "Corrina," and, just as I did in the early days at interim housing, I call Emily so she can listen too.

Si goes on to sing another Taj Mahal song —"Good Morning Miss Brown"—but he fumbles the lyrics, a look of panic passing over his face as his eyes search mine. He recovers, but I can see he is shaken. A table of young men, unfamiliar to me and very drunk, have loudly been requesting that the band play "Copperhead Road."

"Let's try it, Si," Joe says. Simon simply nods and turns his gaze back to the fret board. I am worried for him. It's not a song he's ever really played and certainly not one that he and Joe practiced for this gig. But his shoulders relax and he executes the guitar line flawlessly. Together, the band hits all the breaks and punches in unison, and the cheering and shout-singing from the drunken boys at the back of the room reaches a deafening crescendo.

Later, in the car on the way home, Simon asks for my honest opinion of the gig.

"Well, it seemed like 'Copperhead Road' was a turning point," I say. "You seemed to relax and be more in the moment when you played it."

"Really?" Simon asks. "The only thing that happened there was that I was furious at having to play it. Love Steve Earle, but, man, I hate doing that stupid overplayed song. It's almost as bad as a request for 'Stairway to Heaven' or 'Brown Eyed Girl.' Deadly."

PLAYING LIVE WITH a full band in a noisy bar is difficult. The deafness in Simon's right ear requires him to position himself so that all the other players are on his left side, and for the first time, when he tackles improvised solos, he feels the concrete effect of the brain injury. "The fluidity isn't there," he says, "or the organization of ideas. It's like I can't think in time with the music." He begins a long and arduous process of reeducating himself, listening to previous recordings of the Precious Littles playing live and trying to relearn how his musical mind once worked.

The number one anxiety for Simon in playing before an audience is that the condom catheter and leg-bag setup he uses to deal with his neurogenic bladder might fail and he could publicly wet his pants.

"You'd have to get me offstage right away," he says. "I might not even realize it has happened."

Although there are flashes of his old brilliance, Simon's initial live performances lack consistency, and it takes time for him to build up the stamina to play a full night. He introduces himself to unfamiliar musicians by saying that he is relearning the instrument after a stroke. That he is not the player he used to be. In addition, many of his fellow musicians have moved on to other projects during Simon's long recovery and are too busy to schedule the kind of practice time Simon now requires. He is routinely passed up for gigs and recording opportunities he would have previously been included in. Through it all, Joe remains constant. He is reassuring when Simon has to cancel a gig at the last minute because of illness. He accepts Simon's playing ability on any given day or night and is willing to stand beside him on stage with neither critical commentary nor condescension because, as he says, the only way to get back to playing live is to play live.

"Don't worry," he says after Simon finishes apologizing for a night that had less than Simon's desired level of consistent energy and musicianship. "There's bound to be a few bumps and bruises when you're getting back up on that horse." In this way, their long-standing friendship and playing partnership develops into something deeper, a musical brotherhood that is transformative. One day Simon tells me he doesn't want to have to introduce himself as a player through the accident.

"If somebody asks, I'll tell them, fine," he says. "But I want to get back to the point where my playing speaks for itself."

As his previous level of motivation and determination starts to reassert itself, he takes online courses through Berklee College of Music in Boston and practices every day. He transcribes trumpet solos, reads through his old jazz standard charts, and, most excitingly, starts to write his own songs. "A Pain in the Night" is the first of his originals that he plays for Joe.

"That's a good song, Si," Joe says. "Thumbs up. Big wags."

"It's not that the song redeems the fall," Simon says to me later that night. "But I'm proud of that song. And there is a certain pleasure, or satisfaction, in recognizing I might never have started writing my own songs if not for everything that has happened."

THE PAIN SIMON experiences continues to be a relentless fixture of his days and nights. He knows he needs to continue to strengthen his abdominal and back muscles, but any kind of exertion or exercise triggers an even greater lashing, gripping pain, and he is overwhelmed by the possibility that there is no hope of relief to help him break the vicious cycle. The lack of any kind of exercise has caused him to gain weight, and the wheeling, guitar playing, and waves of pain have rolled his shoulders and spine forward into an uncomfortable hunch. Most worryingly, his arms and shoulders, his basic means of transportation, are often achy and sore now when he transfers.

"Some days, most days, it's just bowel care and pain," he explains to Dr. Willms, his rehab doctor. "There's things I want to do more full-time, playing and teaching music, but I can't get through this to get there."

With the support and guidance of Dr. Willms, we continue to explore various healing therapies: shiatsu, acupuncture, massage, intramuscular stimulation, hydrotherapy, and Botox injections. I, too, make a big decision. I use a portion of the money left to me when my mother's affairs were settled to register in the Classical Pilates Advanced Teacher Training program. Throughout my life my mother bestowed many, many gifts on me; this final one, I believe, will prove to be one of the most fruitful and long-lasting.

The training program, while demanding, is highly flexible, and I am able to integrate it into the schedule Simon and I have established. Marc and Lorna once again support us by coming to visit with Simon on the few occasions when I am required to attend workshops in Vancouver. These workshops provide an important break for me, a breather from the onslaught of Simon's pain; and through a deeper understanding of the Classical Pilates methodology—a methodology meant to decompress joints, increase blood and lymphatic circulation, and promote spinal mobility through a focus on efficient alignment, core strength, and the intrinsic integration of breath and movement—I begin to see, if not an end to Simon's pain, a potential window through it.

There is no big moment, no significant lessening in the neuropathic pain, but gradually, incrementally, something shifts. Simon breathes and stretches. He swims in a pool and allows various practitioners to stick him with needles. He takes vitamin D, magnesium, and fish oil, increases his veggie and protein intake, and cuts down on dairy and gluten. Most importantly, he has things he wants to do. He is no longer content to just play but wants to study, perform, write, produce, and teach music again, and this provides a powerful catalyst for him to continue to move through the pain. Now when a wave of pain hits, he rides it. "It comes; it goes," he says, dismissing the anxious inquiries of friends. In consultation with his GP, he weans himself off of a significant number of the meds he has been on. Listening as he breathes through the pain, he realizes that it is often a communication from the more silent parts of his body: excessive pain in the right hip tends to indicate a flare-up of infection in his right toe; excessive pain in the left hip is a reliable indicator of a urinary tract infection. A general, clamorous increase in pain usually either announces a fever or means he is stressed and verging on overworked and needs some extra sleep or a restful day.

There's a pain in the night calls my name, he writes in his new song. *Calls it so sweet and plain. Now I am all that remains of me.*

{ 29 }

All the Little Things

IN TRUTH I never thought my life would end up here. At night, when I dream, I am up and walking. In the morning, when I open my eyes, the chair is still there, parked beside my bed, a vivid reminder that my body remains in a stasis I can't simply "move" through, a vivid reminder that my legs don't work and that to a certain extent the scope of my life is dictated by where the chair can and can't go.

I never knew what paralysis was like before my accident; I never knew anyone in a chair and so never really understood the possible ongoing health complications of spinal cord injury or the challenges that an inaccessible landscape can present on a daily basis. My injury automatically puts me in the role of raising accessibility awareness in my community. Kara likened it once to how familiar environments were transformed when we were accompanied by a toddler Eli, safe spaces turned suddenly precarious when viewed through the eyes of parents needing to baby-proof: the danger or difficulty is always there, just invisible until your perspective is changed. You don't really see it until you have to.

The mantra at rehab was "Never let the injury define you," and it always kind of irked me. I think it would be more accurate to say "Don't let the limitations of the injury be the defining feature of your life." The reason I make this distinction is because this injury is, for me, a huge part of my identity now. I described it initially this way: When your life shatters, and you survive it, at least you get the option of choosing which broken shards and fragments you want to put back together. Father? Of course. Husband? Yep. Guitar player? All right, let's see how much of that we can reinstate. Fledgling snowboarder? Not so much. What is really, truly important to you, you fight for.

The physical limitations of the injury are also very real. I am far more prone to pain, infection, and fatigue than I was before my accident. Many everyday activities I used to enjoy are more difficult, stressful, or complicated now. How I navigate these limitations also defines who I am, and it is an ongoing process to learn what are the limitations I need to accept and integrate into my life and what are the limitations I need to challenge and push through. And then there is advocating for the elimination of limitations that are externally imposed because of inaccessible environments. These unnecessary limitations are important examples of why accessibility awareness should be an integral component of any community-building project.

I try my best to be honest and unashamed about some of the more difficult aspects of my injuries. I have signed up to an online forum, CareCure, out of Rutgers University. It is an enormous source of practical information and provides a great opportunity for peer support now that I am removed from regular interaction with fellow wheelchair users. Chatting online, I realize I'm not alone in not prioritizing walking as my ultimate number one recovery goal. Bowel, bladder, sexual function— I'd give up the hope of ever walking again just to get control over any one of those.

But I am blessed with a strong and loving family and an inclusive and supportive community that have helped me regain my life, and I look forward to a future filled with music. I am inspired when I hear of the success of my GF Strong friends and fellow inmates. John, the veterinary surgeon who had a high-level incomplete injury, has returned to performing surgeries and has even traveled to Africa to see gorillas in the wild. Although it is a long and difficult process to recover from a spinal cord injury, meaningful recovery can and does happen, and he is a shining example of this. When you go through the process of rehabilitation at a place like GF Strong, and you work so hard to relearn things like eating, shaving, and sitting on a toilet, and you witness others going through similar struggles, you learn to better appreciate and celebrate the trials and triumphs of daily life. The little things (that are not so little any longer) matter a whole lot more now.

{ 30 }

MOVING ON

OCTOBER 30, 2012

M IDNIGHT TONIGHT MARKS twenty-one years from the day we read our handwritten vows to one another, and we spend it overnight in Vancouver. Simon is at Lions Gate Hospital for a surgical procedure to have Botox injected into his bladder. ("I'll have the youngest looking bladder in town," Simon joked with our family doctor. "It will look as good as Cher's face.") The procedure will help him manage his daily catheterizations and ongoing urinary tract infections. While it's not exactly the anniversary trip to Spain we promised ourselves for years, somehow that doesn't diminish our sense of celebration. More than twenty years of marriage: it is an accomplishment to be proud of.

Our sex life is not as straightforward as it once was but, thankfully, it is now nobody's business but our own. While both Simon and I encourage and support more open dialogue around the issue of sexuality and disability, it is also a relief to not have our most intimate selves up for regular discussions with medical professionals or family members. The reality for most couples married twenty years or more is that an evolving love life needs continued commitment to exploration and reinvention. This is maybe more critically true for Simon and me than for most. But the commitment is there, as is the attraction, and we have finally reached a point, post-accident, where we can both have a little fun in figuring out this work-in-progress part of our life.

IN NOVEMBER, JOE and Simon are booked as an acoustic duo for a weekly gig every Thursday at a local restaurant. With the promise of steady work, Simon decides it is time to invest in a good-quality acoustic guitar and finds the perfect fit: a 2009 Santa Cruz.

With this new acoustic duo booking, his guitar students, his songwriting sessions, and the monthly band gig up at Garden Bay, Simon is busy. I am busy too. As the days shorten and the year winds down, I take my second anatomy exam, midway through the year-and-a-half-long Pilates program. I gain fifteen pounds of muscle as the Pilates work continues to help me build and maintain the strength I need to support myself supporting Simon, and I no longer suffer from the low back pain that so often plagued me before and after his accident.

This last six months has marked another transition: both of us have taken on a little more work and responsibility outside of the home. We are no longer exclusively preoccupied with health-related issues and can enjoy the jolts of satisfaction for work well executed, the trivial irritations of being slightly overwhelmed, and the deep pleasure of rest after a day of busyness.

I return home one night after a day of such busyness: an early morning at the Pilates studio, teaching and taking class, rushing home to help Simon midday, and then an afternoon of chores: shopping, recycling, and picking up Simon's prescriptions. It is not yet five o'clock when I get home, but the thin winter light has already leaked from the sky. Simon is in his office chatting on Facebook. Even though we eat early, usually around this time late afternoon, no food prep has been started. I am starving.

"Hey," he says. "What are we having for dinner?"

"Are you kidding me?" I am furious, the kind of furious that only happens when your blood sugar is really low. "What are *you* making? Why haven't *you* started?"

"You're right. I totally flaked," he says, wheeling into the kitchen. "What do you want? Scrambled eggs? Breakfast for dinner?"

I don't really want eggs, but Simon's cooking repertoire is limited, so I grunt in a conditional agreement.

"You're mad," Simon says. He pulls a carton of eggs from the fridge and sprays some oil in the cast-iron frying pan. "I'm sorry."

"I *am* mad," I say, starting to smile. "Hungry-mad."

"What's so funny?" he asks as he cracks the eggs.

"I dreamed about this. Being mad. When you were in ICU. I prayed for the day when I could be righteously pissed off at you."

"Righteously pissed? C'mon, Stan. I'm making you eggs."

"Learn to cook. I need more options than eggs or spaghetti."

"Fine. I will. Now zip it and put some bread in the toaster."

The next day Simon calls me into his office. "Listen to this," he says. He opens a link sent to us by my good friend, and one of Simon's songwriting collaborators, Rachel Rose. It is a live performance by writer and musician Dan Bern, singing about an imagined conversation with God. "Listen." Simon's voice breaks. "It is astounding."

Like the Dylan song "I Was Young When I Left Home" a few years earlier, "God Said No" hits Simon with the emotional weight of a freight train. For three days Simon wrestles with this song, singing and weeping and singing. Unlike the first time, his tears don't frighten me now. Each time he tackles the song, each time it defeats him, he, and his performance, grow more powerful.

"It is a vehicle to navigate regret," he explains, though the song needs no explanation. "What if I hadn't gone into work that day? What if I hadn't stepped back on that scaffold? What if I could go back in time? Do things differently. What if, what if, what if."

The weekly gigs at an Italian restaurant, the Old Boot Eatery, have become more and more popular, and week after week Joe and Simon play to a full house. People line up for the intoxicating combination of wine, crusty squares of bruschetta, spicy homemade sausage, abundant plates of pasta, and good music. The following Thursday, as we drive into town, Simon announces his plan to play "God Said No" live. "I'm a little worried, though," he says. "It might make me cry."

"Cry or not cry," I say, "I think it will be incredible."

He waits until late in the second set, when the crowd has thinned out. Although it is his first time hearing the song, Joe strums along as Simon's voice cuts through the dining-room chatter, and the room falls absolutely silent. Simon makes it through the difficult third verse without choking up, but I notice the woman at the table beside me freely wiping away tears. When he finishes, a man at the back of the room gets to his feet to start the round of applause. My whole body is shaking.

"Whoa," Joe leans in to Simon. "That's quite a song."

PERIODS OF PROGRESS interspersed with "plateau periods" where nothing much changes: that's how Sean, Simon's physiotherapist, described the rehabilitation process to me and Simon in the early days at GF Strong. Over time, he explained, the plateaus would become longer and longer, while the progress would be less remarkable or even noticeable. Still, four and a half years post-accident, progress continues to occur.

Most recently, Simon has decided he will no longer perform with sheet music and pages of lyrics. Before his accident, Simon would never have performed with this type of memory aid but, over the last year, during his weekly gig on Thursday nights, it has proved useful, allowing him to play and sing for three or more hours a night without unduly worrying about forgetting lyrics. But now Simon believes that this "compensation" is hindering his ability to fully emote and perform a song. "If I can't remember the lyrics," he says, "I won't play it." To challenge himself, he learns an intricate gypsy jazz piece entitled "Bistro Fada." It is a song that requires fast, precise, and fluid execution, and it is a significant workout for both his memory and his still sometimes stiff and slow-moving left hand. It is a huge success when he and Joe perform it at the restaurant. Steve, the chef, cheers loudly in the kitchen. Later, he congratulates Joe and Simon. "It was so good, boys," he says, "I kept expecting tables of people to stand up and start smashing dishes!"

There has, however, been no change in Simon's spinal cord injury. Aside from a small increase in spasticity in his left leg and the fluctuations in pain levels, things are as they were the day Simon arrived at GF Strong. In the first year post-rehab Simon would often randomly ask the various medical professionals we encountered what insights they had into cutting-edge research for people with spinal cord injuries. Robotics, one surgeon told us, would likely outpace any biological advances. It was a frustrating answer for Simon, as it implied that the primary goal of recovery was walking. Although it was difficult to accept the prospect of never walking again, Simon has reconciled himself to the continuing presence of a wheelchair in his life and he is quick to assert that walking is at least fourth or fifth on his list of recovery priorities.

In that first year out of rehab he dismissed my suggestion we investigate post-rehab programs like Project Walk, an intensive exercise-based program in Carlsbad, California, whose mission, whatever the level of injury, is to optimize the potential for functional movement. Project Walk began with the idea that if you restrict a normally ambulating person in a wheelchair, give them massive amounts of drugs to dampen the nervous system, and keep them sedentary, in six months to a year they won't be able to walk. This, of course, would be doubly, triply, true of someone with an actual spinal cord injury. The project's basic premise—that an intense exercise program, while not having the power to fully restore sensation and movement, would offer a better picture for the potential for healing—appealed to me. Intense exercise, I proposed to Simon, might also help alleviate pain and would certainly enhance overall health and quality of life.

He wasn't sold. Even the name—Project Walk—irritated and stressed him out. "We don't have the funding for something like that," he argued. "And plus, I have enough on my plate between health issues and getting the left hand working." He was adamant that he didn't want to pursue this type of program.

So, four years later, when I casually assert that I believe he will be up and ambulating one day, I am surprised that he agrees.

We are in Vancouver, driving back to our Holiday Inn room from a gig in a wholly inaccessible venue. Arriving for sound check, we were shocked to find that the "not a problem few stairs leading up to the club" were actually not a few stairs but five steep, long *flights* of stairs. A trio of strong men carried Simon up and down, and Simon is a little giddy at having navigated such an utterly inaccessible environment.

"It makes a whole lot of things feel more possible, you know?"

"Yeah," I say. "But, honestly, I don't know if it will always be necessary to have someone carry you up those kind of steps. I feel like in three to five years' time there is a good chance you'll be up and ambulating."

"I think you might be right," he says. "We'll see."

Reflecting on our brief conversation, I am surprised that this doesn't feel like a momentous breakthrough or epiphany. Somehow we have absorbed the rapid technological advances made both in the study of the central nervous system and in robotics over the past few years. Six months ago I don't think I would have made such a bold statement; nor do I think Simon would have accepted it. Now it simply feels like an inevitable fact.

Some of the most promising research is being done by systems neurophysiologist Miguel Nicolelis. He is picking up the baton in what he terms a long and noble relay race, a journey from the big bang to the human brain and then on to some place and time in the not so distant future that hovers at the cusp of our most speculative imaginings. Nicolelis is a researcher at Duke University whose work investigates the physiological principles that underlie the operation of the vast neural circuitry in the brain. Neural circuits, Nicolelis reminds us in the opening of his book *Beyond Boundaries,* are formed by nerve fibers that in turn are formed by the brain's hundreds of billions of cells. "Such intricate brain networks," he writes, "which dwarf by many orders of magnitude the complexity and connectivity of any electrical, computational, or machine grid ever assembled by humans, allow each individual brain cell, known as a neuron, to establish direct contact and communicate with hundreds or even thousands of its peers ... It is through these immensely interconnected and highly dynamic cellular networks, which are known rather prosaically as neural circuits, that the brain goes about its main business: the production of a multiple of specialized behaviors that collectively define what we usually, and proudly, refer to as 'human nature.'"

Nicolelis sees both the big and the microscopic picture, both the drop of water and the sea, but in the more than two-hundred-year-old battle in neuroscience between localizationists, who believe that distinct brain functions emanate from specialized and segregated areas of the nervous system, and distributionists, who believe that the human brain utilizes populations of multitasking neurons, distributed across multiple locations, to execute specific tasks, Nicolelis firmly situates himself in the latter, more radical camp. But even here, among like-minded maverick thinkers, his vision is so bold and expansive that it is a little frightening. Where others have seen the enigmatic working of the brain's neural circuitry as a highly resourceful and ever-changing map, Nicolelis sees the fine-tuned dynamic choreography of a world-class soccer team or a hunting pack of lions. And he hears the music of the brain. There is the metaphorical song that, to Nicolelis, sounds an awful lot like the nationwide peaceful protest he experienced in April 1984 in Brazil, when a choir of more than a million people demanded fair elections and, through its collective power, was far more influential than any single voice could ever be. But there is also a literal song: in his Duke University research lab, Nicolelis has recorded the sound—the music— produced by ensembles of activated neurons, and he works as a transcriber, an interpreter, of these symphonies of thought composed by the brain. In *Beyond Boundaries* he writes:

> What principles guide the composition and conduction of these neural symphonies? After more than two decades delving into the workings of neural circuits, I have found myself

looking for those principles both outside the brain, beyond the
boundaries that have constrained our biological evolution out
of humble beginnings in stardust, as well as deep inside the
central nervous system, trying to identify and give voice to the
brain's own point of view. Here I propose that, like the universe
that fascinates us so much, the human brain is a relativistic
sculptor; a skillful modeler that fuses neuronal space and time
into an organic continuum responsible for creating all that we
see and feel as reality, including our very sense of being...
in the next decades, by combining such a relativistic view of
the brain with our growing technological ability to listen and
decode even larger and more complex neural symphonies,
neuroscience will eventually push human reach way beyond
the current constraints imposed by our fragile primate bodies
and sense of self.

Nicolelis can make such claims because of the experiments
he has conducted in his lab. By translating the neuronal music
into complex algorithms, he has created a revolutionary neuro-
physiological paradigm, which he has named the *brain-machine
interface,* or BMI. Through the use of a BMI, Nicolelis was
able to successfully teach monkeys to voluntarily control the
movements of various devices—a computer cursor, a robotic
arm or leg—located either close to or very distant from them,
using only their minds. No hands, just thought! As Nicolelis
notes in *Beyond Boundaries,* the success of these experiments
"unleashes a vast array of possibilities for the brain and the
body that could, in the long run, completely change the way we
go about our lives."

Nicolelis's grand and eminently humane vision of the
capacity of the human brain has led him to the verge of developing an exoskeleton to be worn by those who are, either
through trauma or disease, paralyzed. Controlled by a small

computer chip that is powered by, and inserted into, the raw electrical tissue of the brain, these exoskeletons have the potential to restore both sensation and mobility to those who would otherwise be confined to a sedentary life in a wheelchair. There is a sea change happening in our understanding of the central nervous system, and Simon and I, despite our own small-scale daily difficulties, are excited that the possibility of regaining mobility after a severe spinal cord injury is far less remote today than it was even four years ago.

MARCH 2013

THIS IS THE year Simon and I decide to tackle air travel, our most likely destination being his parents' farmhouse in Quebec. But the logistics of air travel continually worry Simon into an aggravated fury.

"No accessible washrooms on the flight," he rants. "I won't even have my wheelchair with me. What if they lose my luggage with all my gear and meds? I've been to Quebec a thousand times. I don't need to go again."

"Is there any place you've always wanted to go but never have that would, in some small way, make the difficulties of air travel worthwhile?"

He thinks for a moment, scratching the silvery stubble on his chin. "New Orleans," he says. "I always wanted to go to New Orleans."

For Marc's birthday, Lorna surprised him with tickets to New Orleans: eight days in mid-March nestled between the high tourist traffic times of Mardi Gras and the Jazz Festival. This particular March she celebrates her sixty-fifth birthday, along with Emily's forty-fifth and Eli's twenty-first, and after a quick consultation we all decide there is no better way to celebrate all these birthdays, in addition to Simon's first foray into air travel, than with a family adventure. So we buy tickets and make plans, and on a Tuesday morning, after much list making, we meet Eli at the Vancouver airport, Emily and Sarah head to international departures at Pearson in Toronto, and Marc and Lorna do the same at Dorval. Late that evening we all meet up in the lobby of the Hotel Monteleone on Royal Street in the heart of the French Quarter. The city, in spring, is draped with garlands of purple and white wisteria.

The vacation is not without its alarming moments—Lorna is aghast when a cockroach the size of a small puppy darts out from beneath her restaurant chair; Eli has a gun pointed at him on Bourbon Street; and, on the way home, the airline temporarily loses Simon's wheelchair in Fort Worth, Texas—but these moments are fleeting and we survive unscathed. We spend our days listening to the buskers who line Royal Street, buying ink prints from the vendors in Jackson Square, and lounging in stone courtyard cafés eating po-boys and oysters on the half-shell. At night we bar-hop along Frenchmen Street, listening to the rich Creole mix of blues, funk, jazz, and Dixieland bands. From the banks of the Mississippi we watch the steamboats pass by, Simon playing guitar and all of us singing Jesse Winchester's "Mississippi You're on My Mind." After a week, a deep family bond has been forged with this beautifully cacophonous city. Although the story of New Orleans is complex and reaches far back into history, we all feel a symbolic, if superficial, kinship. Like us, New Orleans recognizes that trauma and pain are powerful teachers of how to live in the present moment. Our Horses, those running threads that weave through our life and knit us together, the fibers that allow us to find the unexpected reserves of strength and flexibility necessary to navigate, even celebrate, life's inherent chaos and uncertainty—these Horses are ones we find galloping through cobblestone alleyways and holding up traffic on Canal Street. Our Horses: the drive to create more than you consume and a deep and abiding love of good food, good friends, and good family. And good music. Always good music.

JULY 22, 2013

IT IS FIVE years since the day of Simon's accident. "Anniversary" isn't exactly the right word, implying, as it does, celebration. Still, five years feels important, and we need in some small way to honor Simon's courage and resilience and to mark this important milestone in our journey together. We sit on our porch and nosh our way through oversized gourmet takeout burgers and generous portions of our good friend Amy's homemade vanilla ice cream, and then Simon pulls out the new Santa Cruz acoustic.

We have continued to sing regularly. As my voice grows stronger, our repertoire expands and Simon becomes more and more demanding, challenging me to hold notes longer or to sing vocal lines separate from his. Some days it works better than others. But today there will be no holding back. "Ready?" he asks and launches into David Bowie's "Five Years," the song we have been singing together since we were teenagers. We continue with the rest of the *Ziggy Stardust* album: "Moonage Daydream," "Starman," "Suffragette City," even a poorly remembered version of "Rock 'n' Roll Suicide." David Bowie leads to T. Rex leads to the Stones leads, of course, back to the Beatles and raucous versions of "Twist and Shout," "Helter Skelter," and "Why Don't We Do It in the Road?" Sexy songs: jubilant and defiant and playful. It is true that the last five years have felt like an entire lifetime, but it is also equally true that, here, now, as we sing at the top of our lungs at the kitchen table, it doesn't feel so long ago that we were sixteen and on lunch break, dancing around the stacks of albums that lined Simon's sun-drenched sky-blue living room.

Five days later, back on the porch, the dog and I lie on the warm cedar planks in the mid-afternoon sun, the air soft and fragrant with honeysuckle, rose, and the jasmine-like scent that wafts up from the row of Daphne bushes I planted last month. I eavesdrop, as I so often do, on Joe and Simon playing music. Their music today, of all days, counts as a minor miracle to me.

This time five years ago, two things were happening. Simon was being prepped for the controversial surgery to stabilize his spine—controversial because the consensus of the gathered doctors was that there was almost no chance he would survive his injuries. And Joe was on stage at the Islands Folk Festival announcing to the audience that the Precious Littles were a man short, that Simon, his good friend, musical cohort, and producer/guitar player, couldn't be there that day. Today our living room is filled with recording gear, and they are hard at work recording a new CD titled, fittingly, *Good Road Home,* a collection comprising both Joe's and Simon's original songs, songs created in the wake of Si's accident and inspired by their musical sojourn together.

Inside, Simon makes a loud foghorn noise, and he, Joe, and the engineer, David, laugh loudly as they move from living room to kitchen, where we have laid out a spread of finger foods: squares of sharp cheddar, slices of cantaloupe, a bowl of the lime saffron cashews both Simon and Joe love. I've missed both lead-up and punch line, but I imagine Simon is teasing Joe about the glorious honking bass harmonica track Joe played on "Old Fashioned Morphine" earlier that day.

"Don't worry, Joe," Simon says. "I'm just pulling your ponytail. You know I love ya, man."

On the porch, the dog, Paloma, gets up and moves into the shade, settling down with a long, expressive sigh. I am reminded suddenly of Eli's question when I announced I was writing a nonfiction account of Simon's accident: How will it end?

Eli is in Victoria, the first year he hasn't returned home for summer vacation. These are the early days of his being truly independent, an adult, something I know is both terrifying and exhilarating for him. The recording of this CD—the sending of barely weaned songs out into the big wide world—is the start of something new too. In this moment I am content to simply know that our story isn't over. That there is still time yet to listen for music heard so deeply it is barely heard at all. And to know that to that song there is no ending. Only new beginnings.

Brief break over, Joe and Simon return to the living room to tackle the next song: "God Said No." It's one of two covers that has made the list alongside their originals.

From the porch I listen as Simon sings, his voice no longer full of ragged wind and rain but deep and resonant. Powerful. *God said Time, Time belongs to me; Time's my secret weapon, my final advantage. God turned away from the edge of town and I knew I was beaten and that now was all I had. God said no.*

Joe and Simon exhale as the final note fades.

"Playback? Do you want to listen to it?" David asks after a moment of silence. "Or do you want to try that one again?"

"No, I'm good," Simon says. "Let's move on."

AFTERWORD

IT IS IMPORTANT to acknowledge that when telling a story that details a particularly vivid timeframe in the life of a family, there may be multiple versions of shared events. This book is my version. As such, it privileges my feelings and perspective, and because I both lived and wrote it with Simon's constant influence, input, and insight, it tells *our* story—his and mine. Our family fully supports the telling of this story; however, that does not mean there were not some differing opinions about the interpretation or representation of events. Where these conflicts were particularly intense (for example, the wisdom of the decision to give consent for the tracheostomy was long debated), I have attempted to widen my lens to include outside perspectives or at least allude to their existence.

Simon, being Simon, has often expressed something between gratitude and relief that it was he who fell and not someone else—Eli, say, or me. Still, he has always been acutely aware that his accident did not happen to him alone and that the rippling aftershock of his fall spread out among his friends and family, affecting them in profound but less visible and easily comprehensible ways. All of these people formed a chorus of voices in my mind as I wrote this memoir, and while, given the demands of the genre, it would neither be possible nor appropriate for me to have attempted to tell all their stories, those stories are critically important to me and Simon. All these people's communal love and support has buoyed, nourished, and nurtured us in countless ways; we are where we are today because of all of these voices.

The intersection of stories is most complex with our son, Eli. He is an intrinsic part of our story, and at the same time he is on the precipice of forging his own life, as his own person. During Simon's long recovery, Eli has often been very much on his own, navigating a world of "normal" experiences—winning soccer tournaments, writing exams, experiencing first love, applying for university, getting (and hating) a minimum-wage job—while still experiencing the internal fracturing of our nuclear family and everything, really, he had known and relied on as daily life. As much as the early days of rehab at GF Strong and interim housing were slow-going for Simon and me, they gave us time to both grieve and integrate the accident into our life. But it is hard, almost impossible, for a sixteen-year-old to know that the only way to get through sorrow is through the sorrow itself. Too much is happening that is good and terrifying, new and exciting, to slow down for sorrow. It is impossible for this juxtaposition—the internal disharmony of the predictable hopes and worries of a teenager battling the too-soon grief of an adult—not to create conflict, but although it has been challenging at times, Eli continues to work at navigating the perilous emotional place he occasional finds himself in. At twenty-two, he is both still the cornerstone of our story and his own man, writing his own life story with his characteristic conviction, commitment to communication, and an open, generous, and courageous heart.

At my request, Joe Stanton, Gerry Millar, and Marc and (especially) Lorna Paradis wrote extensively for me about their experience of Simon's accident and the years that followed, greatly widening the scope of my view and providing insights I would not have had on my own. Emily Paradis gave me access to her journal entries and private emails, one of which appears in the text and all of which were written from the singular viewpoint of a person whose great intelligence is matched by her deep and abiding compassion. There are more ways than I can list here that the text of *Fallen*, as well as our family, has benefited from Emily's open-minded input. Sister-in-love indeed! All of the abovementioned are gifted writers and storytellers in their own right, and I am honored that they entrusted me with their contributions.

Again, this is especially true for Lorna's contribution. Given our very different personalities (not to mention the social roles of mother and daughter-in-law that we occupy), Lorna and I occasionally had perspectives that clashed or differed. Despite these conflicts, Lorna has unceasingly supported me in my writing of this memoir. And for me, both the act of writing and the attempt to widen my perspective have only deepened my respect, admiration, and love for Lorna, and for our entire extended family.

A great number of important people, facts, and events did not make it into this memoir. Emily's wife, Sarah Fowlie, and my brother, Rob Stanley, figure only as peripheral characters, largely because they could not be as physically present in our lives in British Columbia as other members of the family. In real life they are both superstars. Sarah, foundational source of comfort, nourishment, and wicked humor; Rob, who always asks for very little and gives a great deal—even from a distance these two are central figures in the family support system we rely so heavily upon.

Several holistic health practitioners—Baelay, Orianne, Marianne, Sara, and, in particular, Mary Boulding—have played a pivotal role in Simon's continuing recovery. Mary is a herbalist and shiatsu practitioner; while Simon was in the hospital and rehab, she volunteered to travel once a week to the city, where she would give Eli and me guerrilla shiatsu treatments wherever possible: in a hotel room, on the grass in a park, sitting in a cafeteria chair. She noted, when treating Simon, how strongly his qi, (or energy flow, in eastern medicine), moved through his unmoving, unfeeling legs, and she was a passionate advocate for patience when healing.

In rehab, Simon's new, altered body was alarming, and there was a great deal of fear in knowing, or not knowing, how it could be touched or moved without causing further harm. It takes no great leap of imagination or empathy to understand how it would be possible for a survivor of such catastrophic injuries to want to disconnect and disown a body that was no longer fully functional. To want, as Matthew Sanford articulates in his beautiful memoir, *Waking,* to be disembodied, simply a floating head and shoulders, with relatively little connection to a body that has (by being vulnerable, fragile, mortal, and in pain) betrayed you. But, at that critical moment, we were blessed with Mary. Mary guided Eli and me in simple hands-on shiatsu techniques that we could use to help Simon, a small act but a huge step forward for all three of us in reconnecting with and welcoming Simon's physical body back to daily life.

Mary also brought the considerable power of her combined scientific and intuitive understanding in creating teas, tinctures, formulas, and broths to aid in rebuilding Simon's overall health. The formulas she prepared for me were pivotal in helping me get sleep when I needed to stay grounded, focused, and healthy. In my opinion, Mary is a shining example of what the term "holistic medicine" means, the strength it provides; her healing approach, to the whole body and to the whole family, was a critical factor in supporting Simon, Eli, and me in staying connected within ourselves and to each other.

The role WorkSafeBC has played in helping our family navigate our small personal crisis is also not fully fleshed out in this memoir. I know people do not always feel heard and supported by large government institutions, and it is these examples, when people fall through the bureaucratic cracks and do not receive the compensation they need or feel they deserve, that usually command the most public attention. But it is also important to acknowledge when a system in place does not fail, when it does exactly what it is mandated to do. In our case, that is what happened. While WorkSafeBC's financial support has been paramount in maintaining our domestic stability, its support has gone much further. WorkSafeBC provides medical, vocational, recreational, and psychological support and has continued to champion Simon's achievements throughout his recovery. Simon often jokes that he is a "ward" of WorkSafeBC, but when pressed he will tell you that under Special Care Services director Jennifer Leyen, with her holistic approach, his relationship with WorkSafeBC is one of future-building and empowerment.

LATE ONE NIGHT when Simon was in the ICU, a nurse confided that he often struggled with the work he did. He told me that he often would question the value of the extraordinary measures that were taken in the ICU to delay what only fifty years ago would have been inevitable death. Where was the good in prolonging a life that had no hope for a meaningful recovery? Was it in fact more cruel than compassionate to save

someone only to commit them, and their families, to an ago-
nizingly extended death, their remaining days, however long,
spent in hospital or hospice, navigating various infections and
medical interventions? The implication was that Simon might
well be one of these cases. At the time I was both relieved and
shocked to hear my own horror-thoughts articulated in such a
reasoned and professional manner. "But," the nurse continued, "occasionally you spend days working with someone you
are certain is hopeless"—again, the implication that this was
the case with Simon—"and a few weeks later you meet them in
the hospital hallways, awake and responsive, and then all of it
feels very worthwhile."

The story was told to inspire hope, and although hope for
me at the time was too costly a luxury, it did make me feel marginally better. In the aftermath of Si's accident and through the
process of writing this book, I have come to better appreciate
the kind of difficult decisions that are made daily—hourly—in
the ICU. So, to those of you who took a leap of science (or a
leap of faith in science), to the first responders and paramedics, the ER doctors, the surgeons, and the ICU staff, thank you.
From our limited individual perspectives, and despite the ups
and downs Simon has experienced in his continuing recovery,
your efforts were magnificently, gloriously worth it.

ONE OF THE most startling and breathtaking moments of this
recovery for me was the day that Simon, newly emerging from
his coma and still without spoken words, scribbled the phrase
"A Blank Page for Eli." This message resonated with me for
several different reasons. First, it spoke of and connected to
Simon's thought process as a father. The purity and beauty of
the thought still has the power to instantly bring me to tears.
It is, really, what every parent longs to give their child, isn't it?
A blank page, a clean and unencumbered start, unknown and

infinite possibility. This to me epitomizes the grace of Simon's mind: that in the soupy, mushy, swirly, smudgy thought processes of a mind emerging from the deepest levels of unconsciousness, these were some of the first words, the first thoughts, that surfaced.

This message also resonated with me because, unbeknownst to Simon, it echoes the title of one of my favorite short stories, Isak Dinesen's "The Blank Page," which tells the story of an ancient and once illustrious order of Carmelite nuns who grow flax and weave it into fine linen in the Portuguese mountains. The fine linen is used to grace the wedding beds of the Portuguese nobility, after which a square of the linen bearing the bloodstain that attests to the bride's virginity is returned to the convent, where it is framed, engraved with the name of a bride-princess, and hung upon a great wall. In the midst of this great wall hangs one frame engraved with no name, the linen unstained: a snowy white blank page. The thought of this story is that the most interesting, the most subversive stories often remain untold or reside in the silences of the unwritten page.

Although Simon has made some written contributions to the memoir, the final omission, the blank page of this story, is Simon's own voice. I am lucky. I worked on this memoir literally side by side with Simon (in our separate offices) as he told his story with music: with chords, melodies, and rhythms. That is how he fills his snowy white blank page, and it was the companion soundtrack to the writing of this text. As we do in life, together the words and music form a more complete picture of the whole story. You can check his music out here: http://www.stantonparadis.com/.

ACKNOWLEDGMENTS

THE SUPPORT REQUIRED to write this book began long before I ever set word to page. There were, especially in the first year after Si's accident, so many people that helped out in such profound ways that I had to, in the writing of the memoir, resist the urge to name them all—the cast of characters would have been too huge and unwieldy for any reader to keep track of. But they all deserve to be thanked by name. So, to Lynn Simmons, who delayed the start of a new job to hop on a plane and come live with Eli while he settled into grade eleven: thank you. To John, Colleen, Nate, Jonah, Terra, Cam, Drew, Josh, the Kaizers, and the Reeds (and anyone else who opened their home to Eli): thank you. To Barb and Jer, for holding open the Chez Paradis doors in North Vancouver, and to Lia, for her valiance, unyielding friendship, historical fine-tuning, and general awesomeness: thank you. To Steve and Nikki, for, specifically, the generous loan of the Cardero attic haven while Simon was at GF and the architectural designs for our new accessible living space, and for, more generally, the compassion, intelligence, and artistry you bring to everyday life: thank you.

To Sylvain Brochu, my dance and yoga teacher, whose potent image of a canoe has floated me through many difficult moments of my life: thank you. To my dear friend Rachel Rose: thank you. Rachel is a writer of such depth and scope that her words have reached out and, like the image of the canoe, guided me through unknown and uncharted waters both during the days and nights in the ICU and, later, when she, my writing cohort, midwifed me through the first and most challenging draft of this memoir. To Guido, Sari, and Nadia, the family we choose: thank you. To our extended musical family both pre- and post- accident—Sue L., Sue P., Cam, Patricia, Pat, Kristi, Loewen, Al, Julie, Susann, Ray, Ron and the Garden Bay Pub crew, Joe, Amy, Amanda, Katherine D., David J. T., Karen G., Tom, Jay, and, of course, Sully, Gerry, and Joe—who have and continue to do so much to fill our family life with song: thank you.

To Dave and Lou and the skilled members of the construction crew that worked on our house, many of whom donated supplies and their precious time: thank you for turning an average space into something exceptional.

Over the course of the last six years we have met too many wonderful and supportive health care professionals to list here, but they all deserve thanks for the good work they do. A special thanks to Dr. Jaschinski, our compassionate, considerate, and highly esteemed GP.

There are so many more I could mention by name, but I hope they will forgive me when I include them all in a general, but deeply heartfelt, thank-you to the Sunshine Coast. We are blessed to live in a place so connected and vibrant and so committed to the concept of community. I could call any number of people at 2:00 a.m. in an emergency and they would come, unquestioningly. This is a human gift that, especially in these days of distance and technology, I do not hold lightly.

To Logan, Natalia, Jason and the wonderful John Overell: thank you for sharing such a pivotal part of our journey with us. It is an honor to know you all.

My wonderful agent, Carolyn Swayze, and all the talented folks at Greystone who believed in the idea of this book enough to make it a reality: Thank you!

And, finally, to the beloved members of the Paradis, Fowlie, and Stanley families, I raise my cup and make a toast: You all teach me what family can and should be. Thank you. And to my mother, Kathryn Stanley, a special kind of gratitude: Mom, you taught me all I know about living with an open heart. I owe you my life.

NOTES AND REFERENCES

EPIGRAPH *We need only view*
Niels Stensen, *Discours sur l'Anatomie du Cerveau,* in Stanley Finger, *Origins of Neuroscience: A History of Explorations into Brain Function* (New York: Oxford University Press, 1994), p. 24.

7–8 *The study followed ninety-nine patients*
Joan Didion, *The Year of Magical Thinking* (New York: Alfred A. Knopf, 2005), pp. 94–95.

9–11 *Chapter Two: A Hungry Beast*
Norman Doidge, *The Brain That Changes Itself: Stories of Personal Triumph from the Frontiers of Brain Science* (New York: Viking Penguin, 2007), p. 290; Mark T. Nielsen and Gerard J. Tortora, *Principles of Human Anatomy,* 11th edition (Hoboken, NJ: Wiley, 2007), pp. 612, 613, 616–617; Trevor Powell, *Head Injury: A Practical Guide* (London: Speechmark, 2007), 19–20; Alla A. Vein and Marion L.C. Maat-Schieman, "Famous Russian brains: Historical attempts to understand intelligence," *Brain,* 131 (2008): 583–90, doi: 10.1093/brain/awm326. Powell's *Head Injury,* a very helpful resource book, provides an excellent overview of the internal cerebral process following a head injury as well as a chart which outlines the criteria used to rate patients on the Glasgow Coma Scale.

9 *humans have the highest brain-to-body-mass ratio*
Mice have a comparable brain-to-body-mass ratio while some insects
and birds have an even higher ratio than humans. As an estimate of
intelligence the brain-to-body-mass ratio is not particularly accurate.
The E.Q., or Encephalization Quotient, is a more refined measurement
of intelligence and here, humans, of all the mammals, are at the head
of the pack.

10 *This living tissue comprises roughly 100 billion neurons*
For many years the estimated number of neurons in the brain has been
100 billion. A 2009 study suggests that this number might be closer
to 86 billion. Azevedo FAC, Carvahalho LRB, Grinberg LT, Farfel JM,
Ferretti REL, Leite REP, Jacob Filho W., Lent R., Herculano-Houzel S.,
"Equal numbers of neuronal and nonneuronal cells make the human
brain an isometrically scaled-up primate brain," J Comp Neurol 513
(2009): 532-541.

16 *Sometimes it's coming up roses*
Joe Stanton, "Sometimes You Win," Perf. The Precious Littles, *Sometimes You Win,* Bearwood Music, 2008.

21–25 *Chapter Four: STAT Craniectomy*
Finger, *Origins of Neuroscience,* pp. 4–6. A further exploration of trepanation can be found in the Geoff Bunn's dynamic and fascinating
BBC Radio 4 broadcast *The History of the Brain,* in the first segment, entitled "A Hole in the Head," first broadcast November
2011 (see http://www.bbc.co.uk/programmes/b017b1zd/episodes/
guide). See also James Cooper, Jeffrey V. Rosenfeld, Lynnette Mur-
ray, Yaseen M. Arabi, Andrew R. Davies, Paul D'Urso, et al. for the
DECRA Trial Investigators and the Australian and New Zealand
Intensive Care Society, "Decompressive craniectomy in diffuse
traumatic brain injury," *The New England Journal of Medicine,* 364
(2011): 1493-1502, doi: 10.1056/NEJMoa102077; Stephen Hon-
eybul, Kwok M. Ho, Christopher R. P. Lind, and Grant R. Gillett,
"Decompressive craniectomy for neurotrauma: The limitations of

applying an outcome prediction model," Acta Neurochirurgica:
The European Journal of Neurosurgery, 152 (2010): 959–64, doi:
10.1007/s00701-010-0626-5; Vishal Kakar, Jabir Nagaria, and
Peter John Kirkpatrick, "The current status of decompressive craniectomy," *British Journal of Neurosurgery*, 23, no. 2 (2009): 147–57,
doi:10.1080/02688690902756702; Clemens M. Schirmer, Albert
A. Ackil Jr., and Adel M. Malek, "Decompressive craniectomy,"
Neurocrit Care, 8 (2008): 456–70, doi: 10.1007/s12028-008-
9082-y; and Graham Teasdale and Bryan Jennet, "Assessment of
coma and impaired consciousness," The Lancet, 304, no. 7872
(1974): 81–84, doi:10.1016/S0140-6736(74)91639-0.

33 TED *talk entitled My Stroke of Insight*
Jill Bolte Taylor, "My Stroke of Insight," TED talk (February 2008),
retrieved at http://www.ted.com/talks/jill_bolte_taylor_s_power-
ful_stroke_of_insight.html.

36 *For most of us, there is only the unattended*
T. S. Eliot, *The Four Quartets* (New York: Harcourt, Brace, 1943). This
passage belongs to the third of the four poems, "The Dry Salvages."

50 *pretend you are brave*
Rachel Rose, *Notes on Arrival and Departure* (Toronto: McClelland &
Stewart, 2005), p. 33.

70-73 *Chapter Eleven: Animal Spirits*
Finger, *Origins of Neuroscience*, pp. 16–20, 27; Nielsen and Tortora,
Principles of Human Anatomy, pp. 571, 612, 616–17, 630.

80 *Martha Carson spiritual*
Martha Carson, "You Can't Stand Up Alone," perf. Martha Carson,
Rock-a My Soul, RCA Records, 1957, LP.

86 *Keepers of private notebooks*
Joan Didion, "On Keeping a Notebook," *Slouching Towards Bethlehem*
(New York: Farrar, Straus and Giroux, 1992), pp. 132–33.

97 *In fact I had no idea why*
Didion, *The Year of Magical Thinking*, p. 125.

98 *The pineal gland is small*
Nielsen and Tortora, *Principles of Human Anatomy*, p. 746.

98–99 *René Descartes claimed as the soul's conduit*
Finger, *Origins of Neuroscience*, p. 26.

101 *I buy her book*
Jill B. Taylor, *My Stroke of Insight: A Brain Scientist's Personal Journey* (New York: Viking, 2008).

102–3 *During the Enlightenment*
Bunn, "The Spark of Being." Dr. Bunn's ten-part cultural history of the brain, *A History of the Brain* (BBC Radio 4, 2011), documents how political, philosophical, and cultural mores influenced and were influenced by scientific exploration into brain function over the course of centuries. This paragraph also developed through conversations with Lia Paradis, historian and Simon's cousin. It was she who helped refine my examples of the democratization of knowledge and culture, pointing out the pivotal publications—Newton's *Principia* and Diderot's *Encyclopédie*—that bookended the Age of Reason.

103 *Giovanni Lancisi*
Finger, *Origins of Neuroscience*, p. 387.

120–21 *The two hemispheres of the cerebrum*
Nielsen and Tortora, *Principles of Human Anatomy*, pp. 630-31.

123–24 *a second, more colorful picture*
Condensed and adapted with permission from the author, Dr. Verna Amell; for the full analogy, please see the original source:

Understanding the Brain: A Pictorial Analogy Teaching Guide. Analogy concept: Verna Amell, Ph.D., R. Psychologist. Artwork: Biomedical Communications, University of British Columbia.

Education material reproduced through Education Support at GF
Strong Rehab Centre, 4255 Laurel Street, Vancouver, BC, V5Z 2G9, a
Vancouver Coastal Health Site.

125–26 *Franz Joseph Gall*
Finger, *Origins of Neuroscience*, pp. 32–38. Systems neurophysiologist
   Miguel Nicolelis outlines his belief in phrenology's foundational
   impact in the development of theories of cortical localization in
   his bold and stunning book detailing current, cutting-edge brain
   research *Beyond Boundaries: The New Neuroscience of Connecting
   Brains with Machines—and How It Will Change Our Lives* (New York:
   Henry Holt, 2011); see pp. 6, 36, and 39. Further elaboration of the
   historical relationship between phrenology and cortical localization
   can be found in Geoff Bunn's BBC Radio 4 series *The History of the
   Brain* in the segment entitled "The Beast Within" (November 2011).

126 *Groundbreaking systems neurophysiologist Miguel Nicolelis*
Nicolelis, *Beyond Boundaries*, pp. 5–7.

126 *opposition to this idea has been sporadic but passionate*
Finger, *Origins of Neuroscience*, pp. 35–36, 45, 48, and 53–54; Nicolelis,
   Beyond Boundaries, pp. 26–27, 45–47, and 189.

127 *there is no such strict separation*
Stefan Koelsch, Thomas C. Gunter, Yves V. Cramon, Stefan Zysset,
   Gabriele Lohmann, and Angela D. Friederici, "Bach speaks: A corti-
   cal 'language-network' serves the processing of music," *NeuroImage*,
   17 (2002): 956–66, doi:10.1006/nimg.2002.1154; Anthony Storr,
   Music & the Mind (London: HarperCollins, 1992), pp. 11–12; Steven
   Mithen, *The Singing Neanderthals: The Origin of Music, Language,
   Mind and Body* (Cambridge, MA: Harvard University Press, 2007),
   pp. 5, 26, and 178.

128 *It's a hard way to find out that trouble is real*
Gram Parsons and Bob Buchanan, "Hickory Wind," perf. The Byrds,
   Sweetheart of the Rodeo, Columbia, 1968, LP.

129 *Santiago Ramón y Cajal*
Catherine Y. Wan and Gottfried Schlaug, "Music making as a tool for
    promoting brain plasticity across the life span," *The Neuroscientist*,
    16 (2010): 566-77, doi: 10.1177/1073858410377805.

148 *Jill Bolte Taylor book*
Bolte Taylor, *My Stroke of Insight*.

148 *British publication*
Powell, *Head Injury: A Practical Guide*.

172 *Many would categorize* SCI
H. K. Krueger & Associates for the Rick Hansen Institute, *Spinal Cord
    Injury: Progress in Care and Outcomes in the Last 25 Years* (Delta, BC:
    Rick Hansen Institute, 2011), p. 5.

172–73 *The spinal cord is the multilane pathway*
Nielsen and Tortora, *Principles of Human Anatomy*, pp. 585-90. The
    very informative and comprehensive *Spinal Cord Injury Reference
    Manual* was given to Simon at GF Strong; it consists of material sup-
    plied by various GF Strong employees specific to clients living in
    British Columbia, as well as an extract from the Paralyzed Veterans
    of America's manual: Margaret C. Hammond, *Yes, You Can! A Guide
    to Self-Care for Persons with Spinal Cord Injury*, 3rd edition (Wash-
    ington, DC: Paralyzed Veterans of America, 2000), pp. 1-6.

183–84 *copy of Norman Doidge's* **The Brain That Changes Itself**
Doidge, *The Brain That Changes Itself*, p. xvii. It is impossible to stress
    enough the profound and transformational impact Doidge's book
    had, and continues to have, on my expectations for Simon's recov-
    ery and on my life in general.

198 *It's a hard way to find out trouble is real / In a far away city,
with a far away feel*
Parsons and Buchanan, "Hickory Wind."

229 *I hope my baby's still waiting*

Joe Stanton, "99 Days," perf. The Precious Littles, *Sometimes You Win,* Bearwood Music, 2008, LP.

238 *"The Meaning of Birds"*

Charlie Smith, *Indistinguishable from the Darkness* (New York: W. W. Norton, 1990), pp. 83-84.

258 *I'm going to sing*

Joe Stanton, "Make Me Proud," perf. Joe Stanton and Stanton Paradis, on Joe Stanton, *There You Go,* Old Truck Records, 1997, LP.

258-59 *Music making*

Christian Gaser and Gottfried Schlaug, "Brain structures differ between musicians and non-musicians," *The Journal of Neuroscience,* 23 (2003): 9240-45, retrieved at http://www.jneurosci.org/content/23/27/9240.full; Barbro B. Johansson, "Music and brain plasticity," *European Review,* 14, no. 1 (2006): 49-64, doi: 10.1017/S1062798706000056; Wan and Schlaug, "Music making as a tool for promoting brain plasticity across the life span"; Daniel J. Levitin, *This Is Your Brain on Music: The Science of a Human Obsession* (New York: Penguin, 2006), pp. 83-84, 188, 220-21.

259 *Music might also provide*

Gottfried Schlaug, "Listening to and making music facilitates brain recovery processes," *Annals of the New York Academy of Sciences,* 1169 (2009): 372-73.

261 *I can't go home this a-way...*

Bob Dylan, "I Was Young When I Left Home," perf. by Bob Dylan, *Love and Theft,* Limited Edition, Columbia Records, 2001.

264-65 *The message of chronic pain*

"Pain after spinal cord injury," ch. 22 in *Spinal Cord Injury Reference Manual.*

271 *There's a pain in the night*
Simon Paradis, "Pain in the Night," perf. Stanton Paradis, *Good Road Home,* Cooper Road Studios, 2013.

281–84 *Some of the most promising research*
Nicolelis, Beyond Boundaries, pp. 1–16, especially pp. 5, 7, and 8.

288 *God said Time*
Dan Bern, "God Said No," perf. Dan Bern, *New American Language,* Messenger Records, 2001.

292 *his beautiful memoir, Waking*
Matthew Sanford, *Waking: A Memoir of Trauma and Transcendence* (Emmaus, PA: Rodale, 2006). This philosophical story begins for the author when he is 13; it details the aftermath of a car accident that killed his father and sister and left him paralyzed from the chest down, and follows him into into adulthood and his journey as a yoga practitioner and, eventually, yoga teacher to students of all levels of ability.

FURTHER READING

In addition to Joan Didion's *The Year of Magical Thinking*, Jill Bolte Taylor's *My Stroke of Insight*, Norman Doidge's *The Brain That Changes Itself,* and Matthew Sanford's *Waking: A Memoir of Trauma and Transcendence,* there were several books I read during the most intense period of Simon's rehabilitation and in preparation for beginning this book that were meaningful to me. These were:

Francesco Clark, *Walking Papers: The Accident That Changed My Life, and the Business That Got Me Back on My Feet* (New York: Hyperion, 2010).

Bonnie Klein, *Slow Dance: A Story of Stroke, Love and Disability* (Berkeley, CA: PageMill Press, 1998).

Daniel J. Levitin, *The World in Six Songs: How the Musical Brain Created Human Nature* (Toronto: Penguin, 2008).

Christopher Reeve, *Still Me* (New York: Random House, 1998).